Transitioning to Sustainability Through Research and Development on Ecosystem Services and Biofuels

WORKSHOP SUMMARY

PATRICIA KOSHEL AND KATHLEEN MCALLISTER, *Rapporteurs*

Science and Technology for Sustainability Program
Policy and Global Affairs

NATIONAL RESEARCH COUNCIL
OF THE NATIONAL ACADEMIES

THE NATIONAL ACADEMIES PRESS
Washington, D.C.
www.nap.edu

THE NATIONAL ACADEMIES PRESS 500 Fifth Street, N.W. Washington, DC 20001

NOTICE: The project that is the subject of this report was approved by the Governing Board of the National Research Council, whose members are drawn from the councils of the National Academy of Sciences, the National Academy of Engineering, and the Institute of Medicine. The members of the committee responsible for the report were chosen for their special competences and with regard for appropriate balance.

This workshop was supported by the George and Cynthia Mitchell Endowment for Sustainability Science, the David and Lucile Packard Foundation, the USGS, USFS, NOAA, NASA, USDA, and USEPA. Any opinions, findings, conclusions, or recommendations expressed in this publication are those of the author(s) and do not necessarily reflect the views of the organizations or agencies that provided support for the project.

International Standard Book Number-13: 978-0-309-11982-5
International Standard Book Number-10: 0-309-11982-0

If you would like to request a copy of this report, please call the Science and Technology for Sustainability Program Unit at 202-334-2047 or Email Sustainability@nas.edu.

Additional copies of this report are available from the National Academies Press, 500 Fifth Street, N.W., Lockbox 285, Washington, DC 20055; (800) 624-6242 or (202) 334-3313 (in the Washington metropolitan area); Internet, http://www.nap.edu.

Cover: Photo credit: U.S. Department of Energy. National Renewable Energy Laboratory, PIX13531.

Printed in the United States of America

STEERING COMMITTEE ON TRANSITIONING TO SUSTAINABILITY THROUGH RESEARCH AND DEVELOPMENT ON ECOSYSTEM SERVICES AND BIOFUELS

Pamela Matson (NAS) (Co-Chair), Dean, School of Earth Sciences, Stanford University

James Mahoney (Co-Chair), Former Assistant Secretary of Commerce for Oceans and Atmosphere, US Department of Commerce (retired), Co-Chair of the Roundtable on Science and Technology for Sustainability (until 7/1/2007)

Ann Bartuska, Deputy Chief for Research and Development, US Forest Service

William Clark (NAS), Harvey Brooks Professor of International Science, Public Policy and Human Development Kennedy School of Government, Harvard University

Gregory Crosby, National Program Leader for Sustainable Development, Cooperative State Research, Education, and Extension Service, US Department of Agriculture

Linda Gundersen, Chief Scientist for Geology, US Geological Survey

Alan Hecht, Director of Sustainable Development, Office of Research and Development, US Environmental Protection Agency

Kai Lee, Program Officer, Conservation and Science Program, Packard Foundation

Steve Murawski, Director of Scientific Programs and Chief Science Advisor, National Oceans and Atmospheric Administration

Staff

Patricia Koshel, Senior Program Officer, Science and Technology for Sustainability, The National Academies

Julia Kregenow, Christine Mirzayan Science and Technology Policy Graduate Fellow, The National Academies

Kathleen McAllister, Senior Program Assistant, Science and Technology for Sustainability, The National Academies

Gregory Symmes, Deputy Executive Director, Division on Earth and Life Studies, The National Academies, Director, Roundtable on Science and Technology for Sustainability (until 8/31/ 2007)

Derek Vollmer, Senior Program Associate, Science and Technology for Sustainability, The National Academies

ROUNDTABLE ON SCIENCE AND TECHNOLOGY FOR SUSTAINABILITY

Pamela Matson (Co-Chair), Dean of the School of Earth Sciences and Goldman Professor of Environmental Studies, Department of Geological and Environmental Sciences, Stanford University
Emmy Simmons (Co-Chair), Former Assistant Administrator for Economic Growth, Agriculture, and Trade, US Agency for International
Matt Arnold, Co-founder and Managing Director, Sustainable Finance Ltd.
Arden Bement, Director, National Science Foundation*
Michael Bertolucci, President, Interface Research Corporation
John Carberry, Director of Environmental Technology, DuPont
Leslie Carothers, President, Environmental Law Institute
William Clark, Harvey Brooks Professor of International Science, Public Policy, and Human Development, Harvard University
John Dernbach, Professor of Law, Widener University
Sam Dryden, Managing Director, Wolfensohn & Company
Kathryn Fuller, Chair of the Board of Trustees, Ford Foundation
George Gray, Assistant Administrator, Office of Research and Development, US Environmental Protection Agency*
Hank Habicht, Vice Chairman, Global Environment and Technology Foundation (GETF) and Managing Partner of SAIL Venture Partners
Jeremy Harris, Former Mayor of Honolulu
Rosalyn Hobson, Associate Dean for Graduate Studies, School of Engineering, Virginia Commonwealth University
Jack Kaye, Director, Research and Analysis Program of the Earth-Sun System Division, National Aeronautics and Space Administration*
Gerald Keusch, Assistant Provost, Medical Campus and Associate Dean, School of Public Health, Boston University
Kai Lee, Program Officer, Conservation & Science Program, Packard Foundation
J. Todd Mitchell, Chairman, Board of Directors, Houston Advanced Research Center
Mark Myers, Director, US Geological Survey*
Raymond Orbach, Director, Office of Science, US Department of Energy*
Larry Papay, Former Senior Vice President, Integrated Solutions Sector, SAIC and Senior Vice President and General Manager, Bechtel Technology and Consulting

---

*Denotes Ex-Officio Membership

Merle Pierson, Acting Under Secretary for Research, Education, and Economics, US Department of Agriculture*
Prabhu Pingali, Director, Division of Agricultural and Development Economics, UN Food and Agriculture Organization
Peter Raven, Director, Missouri Botanical Garden and Chair, Division on Earth and Life Studies*
Robert Stephens, International Chair, Multi-State Working Group on Environmental Performance

Staff

Marty Perreault, Director, Roundtable on Science and Technology for Sustainability (As of 9/1/07)
Greg Symmes, Director, Roundtable on Science and Technology for Sustainability (Until 9/1/07)
Pat Koshel, Senior Program Officer
Derek Vollmer, Senior Program Associate
Kathleen McAllister, Senior Program Assistant

*Denotes Ex-Officio Membership

Acknowledgments

This workshop report is the result of efforts by many organizations and people. The workshop steering committee was ably chaired by Roundtable members, Pamela Matson and James Mahoney. Other members of the steering committee included: Ann Bartuska (USFS); Gregory Crosby (USDA); Linda Gundersen (USGS); Alan Hecht (USEPA); Steve Murawski (NOAA); and Roundtable Members, Bill Clark and Kai Lee. Jack Kaye and Woody Turner of NASA, William Chernicoff of the Department of Transportation, and Ann Russell of the NSF also provided valuable support to the workshop.

Gregory Symmes, the former staff director of the Academies' Science and Technology for Sustainability Roundtable, offered valuable guidance to the Academies staff and to the steering committee. We would also like to recognize the contributions made by Julia Kregenow, a Christine Mirzayan Science and Technology Policy Graduate Fellow, Derek Vollmer, and Marty Perreault.

Financial support for the workshop was provided by the US Geological Survey (USGS), the US Forest Service (USFS), the National Oceanic and Atmospheric Administration (NOAA), the National Aeronautics and Space Administration (NASA), the US Department of Agriculture (USDA) and the US Environmental Protection Administration (EPA), the David and Lucile Packard Foundation, and the George and Cynthia Mitchell Endowment for Sustainability Science.

This report has been reviewed in draft form by individuals chosen for their diverse perspectives and technical expertise, in accordance with procedures approved by the National Academies' Report Review Committee. The purpose of this independent review is to provide candid and critical comments that will assist the institution in making its published report as sound as possible and to ensure that the report meets institutional standards for quality and objectivity. The review comments and draft manuscript remain confidential to protect the integrity of the process.

We wish to thank the following individuals for their review of this report: Diana Bauer, US Department of Transportation; Kenneth Cassman, University of Nebraska, Lincoln; Thomas Lovejoy, The H. John Heinz III Center for Science, Economics and the Environment; and Bruce Rodan, US Office of Science and Technology Policy.

Although the reviewers listed above have provided many constructive comments and suggestions, they were not asked to endorse the content of the report, nor did they see the final draft before its release. Responsibility for the final content of this report rests entirely with the authors and the institution.

Contents

1

Introduction and Overview

The National Research Council's Roundtable on Science and Technology for Sustainability hosted ***Transitioning to Sustainability through Research and Development on Ecosystem Services and Biofuels:*** *The National Academies' First Federal Sustainability Research and Development Forum* on October 17th and 18th 2007. The Forum discussed sustainability research and development activities related to ecosystem services and biofuels.[1] The objective of the Forum was to identify research gaps and opportunities for collaboration among federal agencies to meet the challenges to sustainability posed by the need to maintain critical ecosystem services, to support the development of alternatives to conventional fossil fuels, and to manage oceans and coastal areas. The forum focused primarily on federal activities, but included the participation of representatives from the private sector, universities, and nongovernmental organizations.

The Forum's session on ecosystem services featured an introduction defining the concept of ecosystem services and elaborating on some of the most critical research gaps. The gaps included: appropriate theories; multi-scale connections, interactions and tradeoffs; monitoring and indicators; and the design of institutions and policies. Agency presentations emphasized how the results of their R&D activities can lead to more sustainable approaches to natural resource management. Many of the projects discussed addressed the impacts of climate

[1]The Millennium Ecosystem Assessment defines ecosystem services as "the benefits people obtain from ecosystems." (http://www.millenniumassessment.org/en/index.aspx) Biofuels are defined by the U.S. Department of Energy as "liquid fuels and blending components produced from biomass (plant) feedstocks, used primarily for transportation." (http://genomicsgtl.energy.gov/biofuels/transportation.shtml)

variability and land use changes on ecosystem services. Efforts included the development of life-cycle assessment tools for multiple resources, ecological models, and holistic, place based ecosystem analysis. The geographic areas being studied included the Great Plains, California's Central Valley, and regions of Central America.

The session on biofuels addressed many of the themes identified during the session on ecosystem services including the need to examine biofuels holistically—developing frameworks for analyzing environmental and economic impacts associated with various feedstocks and for looking at impacts at different scales. For example, while many cropping decisions are made by individual farmers, their decisions are driven in part by federal and state policies as well as global markets. At the same time, the environmental impacts of production are seen well beyond individual farms with impacts on local streams, watersheds, coastal areas, such as the Gulf of Mexico, and even international changes in land use.

While much of the workshop focused on specific research gaps, participants emphasized that much is already known about the natural science issues associated with ecosystem services and biofuels. The bigger gap is in understanding some of the associated social, economic, political, and behavioral issues. For example, what resource management approaches are most effective and why (and what does not work and why). Is it possible to identify the effects of changing ecosystems on communities and vulnerable people? Who benefits most from current efforts to expand the production of ethanol and other biofuels? Who loses? Many participants emphasized the need to take a holistic approach in looking at the sustainability of biofuels as well as in developing approaches to manage ecosystem services.

A theme throughout the meeting was the importance of strengthening connections between researchers and decision makers—including farmers, fishermen, resource managers, and policy makers. These connections should help shape federal research priorities and assure that the results of this research are used to improve the sustainability of natural systems and local knowledge dissemination—making the results of federal R&D widely available and linking this knowledge to concrete policies and programs at federal, state, and local levels.

BACKGROUND

The National Research Council (NRC) report, *Our Common Journey* (NRC, 1999), described a general strategy for research and development in support of the transition to sustainability—i.e., meeting the needs of present and future generations while substantially reducing poverty and conserving the planet's life support systems. The report stressed the importance of moving beyond "sectoral" approaches to more integrated approaches to sustainability challenges that take into account complex interactions among systems (water, atmosphere/climate, ecosystems, humans, and their institutions). Examples of such integrated challenges to sustainability include those associated with (a) ecosystem services; (b) biofuels; and (c) oceans and coastal resources.

Federal agencies sponsor many programs that support the development of knowledge and the linkage of that knowledge with actions to address sustainability challenges, such as those listed above and described in the NRC report, *Linking Knowledge with Action for Sustainable Development* (NRC, 2006). These programs cover different elements of the research and development continuum, including fundamental research, technology development, and application. Many governmental program activities are tied to specific agency objectives.

Few mechanisms currently exist to identify areas where research and development programs in multiple agencies could be better coordinated to promote the transition to sustainability or to determine critical research gaps and needed analytical tools. Identifying opportunities for cross-agency collaboration is further complicated due to the different mission responsibilities of the agencies supporting these programs.[2] Sustainability issues also cross traditional agency and disciplinary boundaries, demanding contributions from various elements of the research and development continuum. In addition, sustainability presents fundamental challenges in translating the results of research and development programs to actions by natural resource managers, policy makers, and other "customers."

Based on discussions at a scoping session held in March 2007, the Forum steering committee determined that ecosystem services and biofuels would be the major subjects of discussion with more limited discussion of ocean and coastal resources, drawing from the recently released the National Science and Technology Council's Joint Subcommittee on Ocean Science and Technology (JSOST) Ocean Research Priorities Plan.[3]

Prior to the Forum, federal agency representatives were invited to prepare a description of their agency's priorities for work on ecosystem services and biofuels, as well as short descriptions of specific R&D programs on ecosystems services and biofuels supported by their agency. These were shared with the speakers and organizers prior to the Forum, and they provide additional background on programs highlighted during the Forum. Descriptions on the projects presented at the Forum, with minimal edits, are included in Appendixes C & D.

ORGANIZATION OF THE REPORT

This report is limited in scope to the presentations, roundtable discussions, and background documents produced in preparation for the Forum. Chapter 2

---

[2]Several agencies have recently published strategies focused on sustainability including EPA's Sustainability Research Strategy (http://www.epa.gov/sustainability/pdfs/EPA-12057_SRS_R4-1.pdf, accessed on: 3/19/08) and the USGS's Facing Tomorrow's Challenges—U.S. Geological Survey Science in the Decade 2007-2017 (http://www.usgs.gov/science%5Fstrategy/, accessed on: 3/19/08).

[3]Accessed on: 4/2/08, http://ocean.ceq.gov/about/jsost.html

summarizes the presentations and discussions during the ecosystem services part of the Forum, while Chapter 3 summarizes the presentations and discussions during the biofuels session of the meeting. Chapter 4 draws on the presentations made by members of the Roundtable and others describing some of the important cross-cutting, common themes highlighted during the Forum and some of the suggestions for followup activities by the agencies and organizations participating in the Forum as well as by the National Academies.

The appendices to the report include: the Forum agenda; a list of workshop participants; biographical information on the speakers, panelists, organizers of the Forum and members of the Science and Technology for Sustainability Roundtable; a description of federal agency priorities for work on ecosystem services and biofuels; and short descriptions of R&D activities on ecosystems services and biofuels supported by federal agencies.

THE CONTEXT

The Roundtable on Science and Technology for Sustainability was established by The National Academies in 2002 to provide a forum for sharing views, information, and analyses related to sustainability. The Roundtable is co-chaired by Emmy Simmons, Former Assistant Administrator for Economic Growth, Agriculture, and Trade, US Agency for International Development (USAID) and Pamela Matson, Dean of the School of Earth Sciences and Goldman Professor of Environmental Studies, Department of Geological and Environmental Sciences, Stanford University. Members of the Roundtable on Science and Technology for Sustainability include senior decision makers from the U.S. government, industry, academia, and non-profit organizations who are in a position to play a strong role in promoting sustainability. Through its activities, the Roundtable identifies new ways in which science and technology can contribute to sustainability. A list of Roundtable Members is included in the front matter of this workshop summary.

Pamela Matson, Co-chair of the Science and Technology for Sustainability Roundtable, opened the Forum by providing a perspective on sustainability science and technology (Box 1). She explained that sustainability science incorporates a range of science and technology (S&T) (or research and development —R&D) focused on critical challenges: meeting the needs of a still growing population for food and water, energy and shelter while at the same time sustaining and protecting the planet's life support systems—the atmosphere and climate system, water, species, and ecosystems. In particular, she noted that the R&D community can increase public understanding of these challenges, creating new knowledge, tools and approaches to managing these challenges, examining how decisions are made, learning from experience and strengthening mechanisms to promote the use of new knowledge essential to sustainability.

For example, what is the nature of limits—as in carrying capacities and

BOX 1
CHARACTERISTICS AND CORE ISSUES IN SUSTAINABILITY SCIENCE

Characteristics of Sustainability Science and Technology include:

- Use or user-inspired research that contributes to the solution of practical challenges;
- Human-technology-environment interactions;
- Focus on place-based research.

In addition, sustainability R&D often addresses a set of core issues:

- Driving forces, e.g., production-consumption relationships;
- Impacts and limits, such as tipping points;
- Vulnerability and resilience of human-environment systems;
- Incentives for innovation;
- Institutions for governing human-environment systems;
- Valuation of outcomes;
- Designing effective knowledge to action systems.

NOTE: Adapted from Pamela Matson's presentation, Transitioning to Sustainability through Research and Development on Ecosystem Services and Biofuels: The National Academies' First Federal Sustainability Research and Development Forum, October 17, 2007.

tipping points—and the need to recognize the vulnerability and resilience of ecosystems subject to complex and multiple stressors?

The importance of scale, both temporal and geographic, was highlighted. Recognizing a time dimension is a critical aspect of sustainability and sustainability R&D. While there is much that is not known about the characteristics of human and natural systems, it is critical to begin taking steps to protect and manage essential ecosystems and to make more sustainable investment decisions. For example, infrastructure being built today will affect energy supplies, biodiversity, water resources, and countless other assets for years to come. Participants noted that steps need to be taken today to more effectively manage natural systems and thus prevent irreversible harm.

Generally, research on human-environment interactions is conducted in specific ecosystems or geographic regions. However, the challenges to sustainability in China or Africa, or even in diverse regions of the United States, are often different and require different solutions. Nonetheless, there often are some commonalities. For example, driving forces, such as global climate change, may create water scarcity in many areas. Places are connected—what happens in one region can influence what happens in other regions.

2

Ecosystem Services R&D

INTRODUCTION

Although much is known about ecosystem services, a number of research gaps exist, and there are opportunities to strengthen collaboration. One of the major goals of this workshop was to discuss the current work of federal agencies in ecosystem services' R&D related to sustainability while, at the same time, identify opportunities for program/project collaboration. Ecosystem services are the ecological processes that sustain and fulfill human life. General distinctions exist between provisioning, cultural, and regulating ecosystem services. Examples of these services include:

- Provisioning—food, fresh water, fiber, and fuel;
- Cultural—aesthetics, spiritual or educational, recreational betterment of humankind;
- Regulating processes that mitigate floods, purify air, and control agricultural pests.[1]

The public is generally aware of provisioning and cultural ecosystem services, and institutions have been created to manage them, but regulating ecosystem services tend to be less recognizable to the non-technical community.

Many federal agencies have created R&D programs and projects related to ecosystem services and sustainability. Common research foci among the federal agencies' research and development strategies include aspects of climate change,

[1]Accessed on: 4/11/08, Millennium Ecosystem Assessment, http://www.maweb.org/en/index.aspx

life cycle assessments of multiple resources, ecological monitoring, climate modeling, ecosystem management, and holistic, place-based ecosystem analyses.

A goal of the Federal Sustainability R&D Forum was to share information about current R&D in ecosystem services, which was accomplished in part through descriptions of "state of the art" examples by various federal agencies, but also through group discussions during the workshop. A larger goal for the Forum, in addition to learning about each agency's programs and activities, was to identify synergies, gaps, connections, and future opportunities in ecosystem services R&D.

ECOSYSTEM SERVICES: SUSTAINABILITY CHALLENGES AND OPPORTUNITIES

As broadly defined by Steve Carpenter, Professor, University of Wisconsin-Madison, ecosystem services sustain human life. Carpenter began his presentation by defining types of ecosystem services and outlining areas of weakness in the R&D community related to ecosystem services, also noting that significant research efforts already exist. During this session, Carpenter explained that there are often gaps in policies aimed at managing a single natural resource because those policies do not always account for impacts on various other resources with that ecosystem. Based on his research and experience conducting the *Millennium Ecosystem Assessment* (Figure 1) (http://www.millenniumassessment.org/en/index.aspx, Accessed on 4/2/2008), Carpenter elaborated on gaps and opportunities in ecosystem services R&D. He outlined four major gaps: theory, multi-scale connections and interactions, monitoring and indicators, and design of institutions or policies. Although much of the science that is needed to manage ecosystem services exists, more work needs to be done to close the gaps in R&D.

Theory

Carpenter expressed the need for developing indicators that look at where a system is heading to support the theoretical framework surrounding ecosystem services. As the *Millennium Ecosystem Assessment* noted, ecosystem services relate to well-being in different ways, their linkages have different degrees of intensity, and they respond differently to socioeconomic factors. Researchers, for example, should examine the connections between ecosystem services and human well-being to ensure that management is based on human and ecological components. Gaps exist in the understanding of ecological heterogeneity and diversity. Carpenter pointed out that vulnerability/risk assessments need to be thorough and account for potentially drastic changes in ecosystems. Institutions must adapt to the changing needs of both the human and ecological systems.

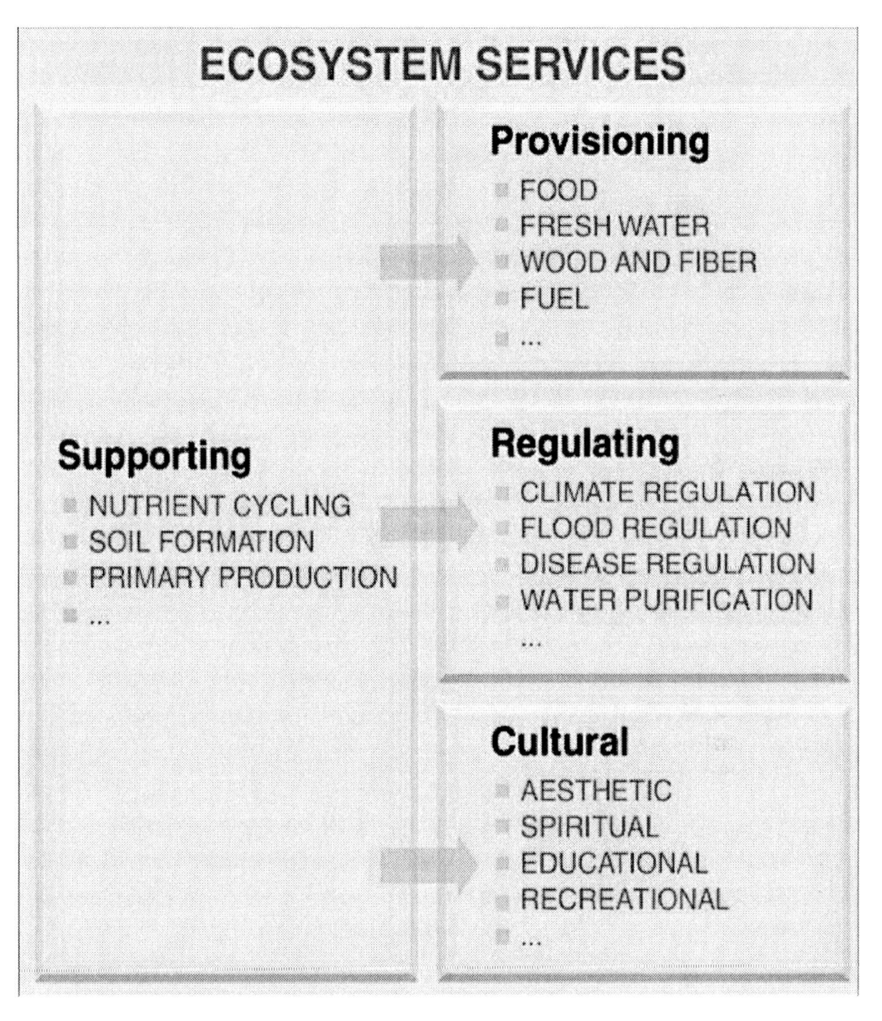

FIGURE 1 *Millennium Ecosystem Assessment*, Available at: http://www.MAweb.org

Multi-Scale Connections, Interactions, and Tradeoffs

Carpenter emphasized the effect of regional processes on local ecosystem services, as well as the impact of local events on regional ecosystem services. Gaps exist in policies aimed at managing a single ecosystem service that do not adequately account for the effects on other ecosystem services. For example, in the central valley of California, many critical ecosystem services are tied to water resources, and as a result, high correlations exist between water resources (quantity, quality, availability) and the ecosystem services in that region. Tradeoffs will be important in recognizing that one management approach cannot be applied to all local-scale ecosystems services. Depending on the unique characteristics of a particular region, tradeoffs may vary. For example, the production of biofuels will have different impacts depending on the characteristics of individual ecosystems, and therefore synergies and conflicts will need to be evaluated by decision makers. Additionally, a given region often provides a "bundle" of ecosystem services; there is a knowledge gap in assessing the bundles of services available (e.g., a plot of forested land with value for its carbon sequestration might also have monetary value to hunters) and also assessing which policies might shift a region from one bundle to another with minimal disruption. Problems exist across local, state, and national levels. There is need for the R&D community to integrate political scales—with institutions and policies at multi-scale levels. There will be progress as NGOs, universities, federal agencies, and the private sector work together and build trusting partnerships.

Monitoring and Indicators

Carpenter noted that despite advances in monitoring technologies, we continue to lack sufficient uninterrupted time series data to accurately reflect socio-ecological dynamics. He also noted that we lack a set of generally agreed upon ecosystem services indicators/metrics which take into account human well being, scale, and direction (where the system is headed). Monitoring and assessment of ecosystem services can be improved by creating a standard set of indicators that include comprehensive time-sensitive information on land cover change, locations and rates of desertification, spatial dynamics of freshwater quantity and quality, stocks, flows, economic value of ecosystem services, and trends in human use of ecosystem services.

Design of Institutions or Policies

A key constraint in identifying effective strategies to create economic incentives supporting ecosystem services is the lack of empirical data documenting the effectiveness of various approaches. Challenges include building institutions to manage use and tradeoffs of ecosystem services, using real-world problem solving to manage services for adaptability and resilience. When looking specifically at federal and local programs, Carpenter noted the gap in coordinating ecosystem management tools across sectors and agencies to address current and future challenges. Currently, society demands a large and increasing amount of natural resources. As a result, it is critical that the management system strengthen the resiliency and adaptability of natural ecosystems.

FEDERAL POLICIES AND RESEARCH PRIORITIES RELATED TO THE ECOSYSTEM SERVICES

Prior to the workshop, various agencies compiled background materials describing their specific research priorities for sustainability and ecosystem services (Appendix C). Although most agencies do not have specific presidential or congressional mandates for sustainability programs, each agency involved in the Forum has developed specific ecosystem management R&D priorities. For example, monitoring and indicators are important R&D foci at the EPA, USGS and USDA (Research, Education, and Economics Directorate). These agencies see a need for a set of indicators that can be applied consistently. There would certainly be advantages in having a manageable set of indicators that each agency could use, along with locally contextualized sets of indicators for specific issues. These same agencies also focus on land use and land management and restoration. Another research focus emphasized in agency research priorities is that of information and decision making. The Forest Service (USFS), EPA, NOAA, and USGS list decision making and policy as key priorities for their R&D strategies related to ecosystem services and sustainability. For example, the USGS "Fisheries: Aquatic and Endangered Resources Program" focuses on the study of aquatic organisms and aquatic habitats. Scientists examine aquatic organisms and their habitats to provide scientific information to natural resource managers and decision makers.

NOAA is one agency that is specifically mandated by Congress to be a leader in protecting, managing, and restoring coastal and marine resources. NOAA plays a vital role in U.S. public health and the nation's economy. A report by the United States' Commission on Ocean Policy led to the *U.S. Ocean Action Plan*, and renewed interest in the world's oceans, their health, and their economic value. On the second day of the Forum, Dan Walker, Senior Policy Analyst at the US Office of Science and Technology Policy, gave a presentation on the Ocean Action Plan. Additionally, NOAA has been moving to an ecosystem approach to management which attempts to achieve a balance among ecological, environmental, and social influences. NOAA's strategy aims to increase public knowledge of ecosystems and sustainability and actively involve the general public as stewards of coastal and marine ecosystems.

The Committee on Environment and Natural Resources (CENR), Ecosystem Services Task Team (ESTT) of the National Science and Technology Council[2] (NSTC) is designed to coordinate and collaborate on research and development across federal agencies. During the workshop, Bruce Rodan, Senior Policy Analyst at the US Office of Science and Technology Policy and Iris Goodman, Co-chair of the ESTT, elaborated on the impact that the ESTT could have on

[2]The NSTC is led by the President's Science Advisor and is responsible for coordinating the federal S&T policy making process; ensuring that S&T policy decisions and programs are consistent with the President's goals, and integrating the President's S&T agenda across agencies.

influencing R&D agendas across agencies. The ESTT acts as a vehicle to bring people together to communicate and coordinate. However, there appears to be an opportunity for the ESTT to be engaged and take full advantage of the programs and new staff at the Forest Service, EPA, USGS and other agencies.

STATE OF THE ART EXAMPLES OF FEDERAL SUSTAINABLITY R&D PROGRAMS/PROJECTS

The following participants described R&D activities related to ecosystem services and sustainability undertaken by the various federal agencies represented at the Forum:

- Ned Euliss, US Geological Survey, *Integrated Landscape Monitoring: Prairie Pilot*
- Mark Nechodom, US Forest Service/Pacific Southwest Research Station, *Alder Springs Fuels Reduction Stewardship Program*
- Iris Goodman, US Environmental Protection Agency, *Willamette-Ecosystem Services Project*
- Dan Kugler, US Department of Agriculture, *Long Term Agroecosystem Research, Education, and Extension*
- Steve Murawski, National Ocean and Atmospheric Administration, *Implementing an Ecosystem Approach to NOAA's Sustainability Mandates*
- Woody Turner, National Aeronautics and Space Administration, *SERVIR: A Platform for the Support of Regional Ecosystem Services for Sustainability*
- Margaret Palmer, University of Maryland Center for Environmental Sciences (National Science Foundation program representative), *CNH: Dynamics of Coupled Natural and Human Systems*

Descriptions of the specific agency projects/programs presented at the Forum are included in Appendix C of this workshop summary and will not be individually summarized here. However, some common research foci arose in multiple presentations, including climate variability, land use, life cycle assessment of multiple resources, ecological modeling, ecosystem management, and holistic, place-based ecosystem analysis.

Major R&D efforts related to sustainability and ecosystem services vary in scale and approach. Like biofuels R&D, place-based ecosystem services research is increasingly important. Although research findings may not always be applicable from one place to another, some important trends can be observed. Many federal R&D programs look at "bundles" of ecosystem services using a place-based approach. For example, the U.S. Geological Survey has partnered with the U.S. Department of Agriculture and other agencies in the U.S. Department of the

Interior to conduct research across the Northern Great Plains region and develop a methodology to quantify the delivery of multiple ecosystem services. The project monitors services such as: floodwater storage, biodiversity, sediments and nutrient loading, and assesses the potential for carbon sequestration and greenhouse gas benefits.

Some R&D is focused on the effects of land use changes on various ecosystem services. Understanding how ecosystems services are impacted and being able to adapt management strategies is vital to long term sustainability.

A number of programs are examining the potential impact of climate variability—especially droughts—on ecosystem services. Other agencies, such as the USFS, are implementing a systems' analysis to look at the benefits and the impacts associated with harvesting biofuels and removing wood and excess biomass[3] to reduce wildfire intensity and reduce greenhouse gas emissions. One project, known as the Alder Springs Fuels Reduction Stewardship Program, suggests that it is possible to achieve ecological objectives that is, fuels' treatments and wildfire emissions or wildfire change, while at the same time provide products such as sawn logs and biomass for energy production. The Alder Springs project uses improved land management techniques to produce biomass for energy production; a project that can offset fossil fuel use. The project also conducts life cycle assessments to examine tradeoffs in habitat, carbon, and watershed values and evaluate the impact of abstracting biofuels from the landscape for the purpose of reducing wildfire impacts.

Various federal agencies are developing ecological models to help decision-makers address issues related to sustainability. For example, NASA uses earth observation technologies to forecast environmental disasters in Central America. NASA has chosen to conduct R&D in this region based on its high biological diversity, high cultural diversity, and high degree of climate variability. Jointly developed with CATHALAC, (Water Center for the Humid Tropics of Latin America and the Caribbean) an international organization based in the Republic of Panama, the regional monitoring system dubbed SERVIR (a Spanish acronym for Regional Visualization and Monitoring System) provided weather predictions for both hurricanes Dean and Felix in August and September 2007. Overall, SERVIR aims to build capacity within the region. The project uses earth observation technologies and prediction functions, which act as decision support tools for protection against natural disasters. By using these prediction technologies for disasters such as landslides, forest fires, floods, and deforestation, decision makers can decide when/if an evacuation is needed and assess what type of natural disaster may occur in the near future. As part of that initiative, SERVIR manag-

[3]Fuel treatment is the process by which wood and excess biomass are removed from a forest to reduce the intensity of wildfires and maintain forest health. *Source: The Sun Grant BioWeb*, http://bioweb.sungrant.org/Home (Accessed on: 4/2/08).

ers educate local weather forecasters about using SERVIR models as forecasting tools. SERVIR also plans to expand into other sectors, specifically agriculture.

The EPA has modeled the Willamette River in Oregon to determine how to take advantage of the river's natural cooling processes to correct the region's natural thermal discharges, thus improving the viability of the region's fisheries.[4] Using the natural cooling processes is cost-effective, and it enables managers to allocate additional money for the restoration of other ecosystem services.

Other federal programs look at the way in which sustainability and economic feasibility are integrally linked. For example, the Research, Education, and Extension Directorate (REE) at the USDA focuses on creating sustainability research programs. Through their proposed long term agro-ecosystem research, education, and extension program, LTAR-EE, the USDA plans to study, design, manage, evaluate and understand hybrid environmental systems with agricultural, natural, and human components.

Still other federal programs have focused their research and development efforts on studying particular management approaches. For example, NOAA assesses alternative approaches to ecosystem management by looking at the development of broad, stakeholder-based governance systems, the conservation of essential components of the ecosystem, and the conservation of essential ecosystem processes. More specifically, this requires evaluating feedback effects and carrying capacity, as well as accounting for climate variability and numerous other variables.

The National Science Foundation (NSF) and the USDA have partnered to form the Dynamics of Coupled Natural and Human Ecosystems (CNH). This program links social scientists and political scientists to examine ecosystem services through a holistic perspective. For example, CNH supports a project, in which researchers, including anthropologists and environmental scientists, at the University of Hawaii, are examining how spiritual beliefs inform farming practices and, in turn, biodiversity in the Amazonian forest in Brazil. This region of Brazil has multiple ethnic groups with varying levels of focus on cosmology. In general, CNH explores interdisciplinary research foci that are central to R&D to enhance sustainability.

Many agencies and workshop participants highlighted the promising strategy of assessing and managing ecosystem services at a place-based level. Although place-based R&D is very useful for local ecosystems, many participants acknowledged the need to explore ways to scale up these research efforts. Program managers could apply what is known about local ecosystem services' management practices to more regional, national, and international levels. Exploring innovative land use management strategies can lead to better use of natural resources and thus more efficient/cost-effective provision of ecosystem services. Climate vari-

[4]This project is an academic grant funded by the EPA through the Collaborative Science and Technology Network for Sustainability (CNS) program.

ability is an important component of ecosystem services' R&D and sustainability. Exploring climate variability through ecological modeling can help to conserve some of the earth's most vital natural resources.

ECOSYSTEMS SERVICES AND SUSTAINABILITY R&D: GAPS AND OPPORTUNITIES FOR INTEGRATION, COORDINATION, PARTNERSHIPS

Bill Clark (Harvard University), Ann Bartuska (USFS), and Sara Scherr (EcoAgriculture Partners) led the introductory panel discussion on ecosystem services gaps and opportunities. Based on roundtable discussions and the state-of-the-art examples presented earlier in the workshop, the panelists asked participants to consider the following issues during breakout discussions, detailed in Box 1.

Breakout sessions were organized for workshop participants to foster discussion of the issues, and each breakout group presented a summary of the discussion.

BOX 1

Set of Common Gaps and Opportiunities for Collaboration in Ecosystem Services R&D Related to Sustainability

- How can we determine what programs/projects will be helpful in building more complementarities among federal agencies?
- How can we acquire more user-driven perspectives to complement the researcher-driven perspectives that were evident in each of the state-of-the-art examples presented by the federal agencies?
- How can we determine the type of indicators that fit into user-driven evaluations? How can we determine what works and what does not?
- How can we create the institutional capacity to actually get the job done? If fundamental R&D is coupled with practical problem solving in the field, how can we determine the types of institutions needed to facilitate connections?
- How can we "Scale Up" the place-based research to apply it more broadly? There is a need to learn how to effectively combine an ecosystem services' perspective with decision making. That decision-making capacity needs to be broadly applicable throughout the nation's ecosystems, land and sea systems, and internationally, in order to move beyond a place-based, one time approach.

NOTE: Before breakout discussions, speakers from the panel on Ecosystem Services R&D Gaps, Opportunities for Integration, Coordination, and Partnerships asked workshop participants to consider these five issues.

ECOSYSTEM SERVICES AND SUSTAINABILITY R&D: SUMMARY OR ROUNDTABLE PANEL DISCUSSIONS

The day's presentations spurred discussion of lessons learned from the past 10 years of ecosystem management studies with increasing attention to expanding management efforts to focus on more inclusive sets of services rather than one or two outputs. The issue of scale of R&D programs continued to come up during workshop discussions. The importance of linking knowledge to action was discussed and was exemplified in various federal examples presented during the workshop, but many participants suggested that there is still work to be done in this area. In particular, regional centers of excellence were highlighted as mechanisms for linking knowledge about ecosystem services research to users and, ultimately, action.

Regional centers of excellence could be populated not just by federal agencies, but also by universities and state and local agencies. The centers could address issues and provide tools at the local level dealing with sustainability issues and combine resources to create these programs. Some participants noted that these research strategies are more effective when they are based on user perspectives and needs. Local-user needs must be understood in order to increase user engagement. If much of the R&D is locally contextualized, other agencies can learn from the various mechanisms for regional management already in use within the EPA and the USFS. Each agency seems to have somewhat different models for this strategy, but they have demonstrated the institutional capacity to achieve success through regionally-based programs. This strategy certainly has much promise in achieving larger R&D goals.

At the federal level, the Ecosystem Services Task Team (ESTT) could provide a link from regional research to larger R&D goals. The ESTT offers an interagency way to collaborate. Many workshop participants see the ESTT as a mechanism strengthening R&D across agencies. Once revitalized, the ESTT could foster more interagency cooperation and perhaps provide a forum to coordinate experiences and knowledge among federal agencies and other R&D organizations.

Many workshop participants discussed the importance of making the concept of ecosystem services "real" through education and communication. Participants also noted the importance of institutions that move science into practice. An appropriate set of ecosystem services' indicators could effectively take science and make it broadly understandable so that the general public will accept them. For example, indicators for water resources might ask: Is this water potable? Is it swimmable? Is it usable on another level? When the public sees this set of indicators, they would be more likely to understand the impacts of their own actions, and what steps they could take to make their community more sustainable. Development of a common set of indicators/metrics across all federal agencies to use for ecosystem services would be a helpful resource. The indicators, of course, would need to accommodate regional variation.

Once the concept of ecosystem services is more broadly understood, it will be important to define the economic value of those services. Decision support tools should be developed specifically for the financial community. By developing indicators, stakeholders may be able to determine how ecosystem services could best be marketed. Additionally, investors should know what the science is, what the decisions are, how the information flows, and what the associated risks and uncertainties are. The valuation of ecosystem services puts a price on natural resource consumption and damage and introduces this concept to a market-driven economy.

A number of participants noted that there is a lack of understanding of the social and behavioral issues surrounding ecosystem services R&D. In the Chesapeake Bay, oyster habitats were examined initially in isolation, but now scientists are looking more broadly at biophysical and socio-cultural issues as well. Unlike traditional socio-cultural impact studies, the cultural analysis conducted by researchers at the University of Maryland asked what oysters meant to a wider range of stakeholders: scientists, restaurant owners, the general public, environmentalists, etc. The study hypothesized that oysters are symbolic for many people because they represent the ecology, economy and culture of the Chesapeake Bay. In this case, stakeholders can provide a wealth of knowledge to the discussion on how to restore oyster habitats.[5] Engagement of the political science community is also vital to overall R&D strategies because of the importance of the policy context within which decisions are made.

Ecosystems services are sometimes cryptic but also very important. Public policy should not ignore the importance of ecosystem services but rather highlight ecosystem services' examples so that the general public is aware of both their impacts on available services, as well as the tradeoffs of maximizing the value of one ecosystem service over another. For example, biofuels from corn, reduce human dependency on fossil fuels, but have negative impacts on water and soils. Many participants acknowledged that incentives and disincentives for collaboration exist between the federal government, NGOs, universities, and industry. Participants noted that these incentives and disincentives should be clarified and plans for collaboration should be discussed.

---

[5] Oyster Restoration in the Chesapeake Bay, A Cultural & Socioeconomic Assessment, March 2008, Non-Native Oyster EIS Executive Summary

3

Biofuels R&D

INTRODUCTION

The United States biofuel industry is growing dramatically with over 90 million acres of corn planted in 2007; a 15 percent increase from 2006, with a large share being used to produce corn based ethanol. There are currently at least 127 corn based ethanol refineries with many more scheduled to come on line over the next few years. At the same time, major R&D efforts are underway to develop commercial scale technologies for producing cellulosic biofuels. While nearly half of U.S. gasoline is blended with at least some ethanol, less than 5 percent of the domestic fuel demand is currently met by ethanol.

This growth in the production of biofuels feedstocks and in the construction of refineries has been stimulated in large part by federal policies, most importantly a 51 cent per gallon subsidy. Support for expanding biofuels production and use has been driven by the need for the U.S. to become more energy independent and to improve long term energy security. It is also driven by the desire to stimulate rural economies and support US farm interests. At the same time, biofuels are seen as a possible way to reduce emissions of greenhouse gases and thus to address the issue of global climate change.

The current administration has made biofuels a centerpiece of its energy policy. In the 2007 State of the Union Address, President Bush committed to expanding U.S. biofuel production and use by seven times current levels—35 billion gallons per year in 10 years. This would lead to a 15 percent reduction in the amount of gasoline which would otherwise be consumed in 2017. This proposed standard for renewable and alternative fuels was called for as part of the "Twenty in Ten" plan, a goal for this country to reduce our gasoline usage by 20 percent in ten years. At the time of the Forum, other ambitious energy goals were being debated.[1] These quantitative goals are an important force driving not only research and development expenditures, but also the market for biofuels.

One goal of the Federal Forum was to share information about current R&D in biofuels, which was accomplished in part through descriptions of state-of-the-art examples by different agencies, but also through group discussions. In addition to learning about existing and proposed federal activities, Forum attendees were encouraged to identify linkages and gaps in biofuels R&D, and to begin the discussion of how to take advantage of possible synergies.

BIOFUELS: SUSTAINABILITY CHALLENGES AND OPPORTUNITIES

Dan Kammen, Director of the Renewable and Appropriate Energy Laboratory at the University of California at Berkeley, opened the biofuels session by describing some of the emerging issues associated with biofuels and the implications of biofuels production and use for sustainability. He focused on the need to consider biofuels as part of a broad energy policy, to take a holistic view; to recognize that corn based ethanol is not likely to be a viable long term solution to increasing energy independence but a short term transitional fuel. He emphasized the need to use a portfolio approach to meeting America's long term energy needs recognizing that increasing supplies of conventional and non-conventional energy and significantly improving energy efficiency must be part of the overall strategy.

Kammen talked about a number of different approaches that could be used to assess the costs and benefits of biofuels and to guide future energy policies and energy investments. He noted that one tool is the current federal renewable fuel standard. However, he suggested that a low carbon fuel standard or a sustainable fuels standard based on expanded life cycle assessment tools is likely to be a better means of assessing alternative energy choices. He discussed the need to address the effects of biofuels production on land use changes as well as a broad range of environmental effects beyond any potential reduction in greenhouse gas emissions such as water quality and quantity, air pollution, soil erosion and sedimentation, and biodiversity loss. In addition, Kammen highlighted some of the social and economic effects both domestically and internationally. For example, the shift of corn-based biofuels has led to increases in prices of food such as corn-based products and meat and diary products. Diversion of land from soybeans to corn has shifted production to other countries where production practices may be damaging to the environment. Kammen suggested that any assessment of biofuels must take a *holistic* view, looking at all benefits and costs. For example, we need

[1]The Energy Independence and Security Act (EISA) was passed in December 2007, setting a goal of 36 billion gallons of biofuel by 2022.

to study the direct and indirect *land use* effects. To fully understand such effects, we need *metrics* (indicators) to assess these factors.

While much attention has been focused on growing feedstocks, less attention has been paid to the production, distribution, and use. Current refinery processes are not particularly efficient and can have serious and wide ranging negative environmental effects. For example, some processing facilities are fueled by coal. These facilities can operate much more sustainably if they are be fueled by alternative energy sources such as wood wastes or other agricultural wastes.

As Kammen discussed, corn-based ethanol is currently at the center of the US biofuel industry, but it is not considered the best long term biofuel. The estimated net energy benefits of corn-based ethanol are minimal and the associated environmental costs are high. Cellulosic-based fuels seem more promising but facilities do not yet exist for large-scale commercial production. Many of the potential cellulosic feedstocks are not expected to require large amounts of new agricultural land and will not affect prices for food and fiber. Some sources could even use zero agricultural land, including algae, off-season crops, or prairie grasses. The wide variety of potential biomass feedstocks was a recurring theme throughout the forum and is discussed more below.

Kammen discussed an important benchmark for consideration—*How much land* would be needed to meet the U.S. energy demand with biofuels? Standard corn grain ethanol requires a great deal of agricultural land, and to meet the entire U.S. demand for liquid fuels would require more than the country's entire land area. Even with improvements in yields and processing efficiency, corn-based ethanol would still require a substantial share of the country's cropland.

Kammen emphasized the need to supplement increases in biofuel production with increases in vehicle efficiency, in many cases using existing technologies. For example, existing technologies could increase vehicle efficiency by 2.5 times reducing transportation fuel needs by more than half. Recognizing trade-offs and options is an important part of any energy strategy. For example, if the entire U.S. corn crop were used to make ethanol, it would displace less gasoline than would raising fleet fuel economy by five miles per gallon. In light of this, discussants suggested that improved fuel efficiency must be part of any long term energy strategy.

FEDERAL POLICIES AND RESEARCH PRIORITIES RELATED TO BIOFUELS

John Mizroch, the Principal Deputy Assistant Secretary in the Office of Energy Efficiency and Renewable Energy at the Department of Energy, discussed the current federal landscape for biofuels. He discussed the new federal mandate for biofuels, DOE's R&D biofuels' priorities and the federal Biomass R&D Board. He also noted the importance of recognizing the international dimension of biofuel production.

The administration's "Twenty in Ten" initiative has set of goal of reducing U.S. gasoline consumption by 20 percent in 10 years, increasing ethanol production to 35 billion gallons by 2017, and by 2030 increase production to 60 billion gallons a year. In order to meet these goals, DOE's R&D programs focus on increasing the range of feedstocks available for fuels and reducing the costs of converting feedstocks to fuels. Recognizing that cellulosic feedstocks will likely be the primary source of ethanol in the future, DOE has committed about $1 billion to cellulosic ethanol production and R&D, which it expects to be matched by private sector funding. DOE expects to support the construction of 16 cellulosic ethanol plants with a least six built at commercial scales.

Much of the federal research on biofuels is coordinated through the Biomass R&D Board chaired by DOE and USDA, which also includes members from the Department of Interior, DOT, and the EPA. The USGS and EPA fund specific R&D programs looking at some of the sustainability aspects of biofuel production and use. For example, the USGS is assessing the impacts of biofuel production on biota, water and land including the effects of returning fallow lands to agricultural production on water quality, native plants, migratory birds, and wildlife. Complementary work is being done by the EPA looking at the effects of the entire biofuel system on natural and manmade systems.

Alternative fuels are an *international* issue, though much of the discussion in the United States tends to focus on the domestic aspects. The world is becoming rapidly urbanized with increasing demands for energy and increased requirements for petroleum and other liquid fuels. Much of this demand is driven by the dramatic increase in car ownership in countries such as China. Many countries are beginning to develop or expand their own biofuels production capacity.

STATE-OF-THE-ART EXAMPLES OF SUSTAINABLITY R&D

The following forum participants described examples of the biofuels R&D being supported by federal agencies:

- Jeff Steiner, US Department of Agriculture (USDA), *REAP: Renewable Energy Assessment Project*
- Richard Alexander, US Geological Survey (USGS), *USGS Research on Biofuels Sustainability*
- Randy Bruins, Environmental Protection Agency (EPA), *Future Midwestern Landscapes*
- Marcia Patton-Mallory, US Forest Service (USFS), *Bioenergy from Forests-Moving Science to Practice Issues of Sustainability*
- William Chernicoff, US Department of Transportation (USDOT), *Integrated Mapping of Biofuels Feedstock Production and Transportation for Systems Visualization and Optimization*

- Jacques Beaudry-Losique, Department of Energy (DOE), *Sustainability and Biofuels Production: A DOE Perspective*

Summary descriptions of the projects presented are included in Appendix D of this workshop report. The efforts are dominated by the programs in DOE and USDA but other agencies also have R&D programs focused on sustainability and biofuels. Some of the common issues addressed in agency R&D efforts included: evaluations of alternative feedstocks, the effects of production on ecosystem services, transportation requirements, and developing methodologies to assess the economic, environmental, social, and technical aspects associated with fuel choices.

An increasing variety of feedstocks are now being considered for biofuel production. Corn-based ethanol has received the most attention because it commands the majority of the current biofuels market and additional opportunities to improve cropping practices and conversion efficiencies. However, many other sources of biomass can be used. These options range from agricultural crops (including sorghum, soybeans, and sugar) to agricultural residues (e.g., corn stover or wheat straw) to energy crops (such as switchgrass) to forest residues (e.g., tree thinnings, logging scraps, sawdust) to wastes (including recycled grease, garbage, and manure), to algae. These are not mutually exclusive categories. For example, Marcia Patton-Mallory (USFS) described a feedstock that is both a forest residue and a waste—the wood that builds up on the forest floor because natural forest fires are often suppressed. This wood buildup raises the likelihood and severity of future forest fires. If it can be collected and used as fuel, it is basically a win-win situation. Since finding ways to dispose of waste is also a sustainability challenge, turning waste into an energy source is an attractive option. At the same time, corn production can continue to provide a feedstock for ethanol and the wastes, corn stover, can provide a feedstock for cellulosic ethanol.

While there are currently no commercial scale cellulosic plants, some 16 cellulosic ethanol production facilities are expected to be operable in the U.S. by 2010. Jacques Beaudry-Losique (DOE) described DOE's new research activities focused on the commercialization of cellulosic ethanol. He described some of the advantages of cellulosic biomass over corn as an ethanol feedstock. He noted that cellulosic feedstocks do not compete with food crops and require less fertilizer. Furthermore, cellulosic-based ethanol releases substantially lower carbon dioxide (CO_2) than corn-based ethanol.

There are additional forest resources that are available or underutilized that could be used to create biofuels. Marcia-Patton Mallory (USFS) talked about the potential role of forest resources as a source of bioenergy. She noted that almost 50 percent of US renewable energy supplies are from biomass, mostly used for heat and power. Many of these forest resources are wastes and need to be cleared as part of a sustainable forest management strategy reducing the potential for forest fires and creating healthier forests. Dr. Patton Mallory emphasized the need for an integrated or holistic approach to looking at a variety of bioenergy feedstocks and for evaluating the sustainability implications of different approaches.

Several participants stressed the need to understand the effects of growing various feedstocks on water, land, air, and soil. Randy Bruins (EPA) discussed a new study being conducted by the EPA which examines how changes in land use and the development of biofuel production facilities will potentially influence ecosystem services. The project is expected to create a decision tool kit which can assist producers and policy makers in understanding the implications of biofuel investments on critical ecosystem services including air and water quality, local hydrology, natural areas, and wildlife.

Corn stover is a crucial natural fertilizer that helps maintain the soil's organic carbon content as well as other nutrients such as nitrogen, phosphorous, and potassium. Jeff Steiner (USDA) focused on potential changes in soil quality from the use of corn stover as a biofuel feedstock He noted that the stover is usually left on the soil to maintain soil productivity and reduce erosion. If the stover is harvested, soil quality will suffer reducing subsequent crop yields. Determining the amount of stover that must be retained on the soil and that which can be used as a fuel is a big research challenge.

Growing feedstocks for biofuels requires a lot of water input, but the water output also significantly affects the surrounding aquatic ecosystems. Richard Alexander (USGS) focused on the relationship between water and biofuels. When considering the role of water in an ecosystem, both quantity and quality are important properties. Runoff from agricultural land delivers nutrients from fertilizers into streams which in turn impact downstream areas. For example, agriculture is the main source of nitrogen and phosphorous in the Mississippi River Basin and in the Gulf of Mexico waters. This excessive buildup of nutrients has contributed to the high dissolved oxygen levels killing many aquatic organisms.

Steve Parker (National Academies) similarly highlighted the important role of water in his description of a recent study by the NRC's Water Science and Technology Board on *Water Implications of Biofuels Production in the United States* (NRC, 2007). He discussed the effects of the expanding biofuels industry on water, including both crop production needs—also mentioned by Mr. Alexander—and ethanol processing, which is water-intensive. He described irrigating corn crops as the dominant "worst case" in water resource use for crop production. Secondary, but not insignificant, are the water requirements and impacts from the production facilities themselves.

Transportation of fuels between where they are produced and used is an important ingredient in the calculus of their sustainability. If biofuel production is significantly increased, how will we meet the extra transportation need? It is important to consider the method and the distance of transport. The method is a key question for ethanol, for example, which typically must travel by truck, waterway, or rail. Ethanol is miscible with water and thus subject to mixing with pipeline impurities which do not mix with petroleum or natural gas fuels, and thus cannot easily be transported through existing petroleum pipelines.

In addition to transport method, distance is also a key consideration. William Chernicoff (DOT) pointed out that the further a fuel has to be transported, the higher its final cost, and the less sustainable it is. Emphasizing the important role his organization plays in energy sustainability, he added that two-thirds of energy in the United States currently moves through the DOT's transportation system. Furthermore, there may not be sufficient capacity in the locations where it is needed. If the additional biofuel load on the existing transportation system increases overall traffic congestion, this resulting loss of efficiency increases the waste of all fuels and again decreases the sustainability of a fuel.

Optimizing biofuel feedstocks to a regional climate and soil condition and then using that fuel locally would minimize the costs of transport required and often improve sustainability. Moreover, maintaining a diverse portfolio of fuel sources could make the market more robust, insulating it from potentially devastating effects of occasional crop failures, for example. The importance of such a holistic, big-picture perspective of the problem was echoed by many participants in a variety of contexts throughout the forum.

Many participants discussed using a holistic approach to examine biofuels and sustainability. A holistic approach also requires considering the lifecycle of a fuel. This includes the broad ecosystem perspective discussed above, which means taking into account a feedstock's effects on all of the resources of an ecosystem. A lifecycle analysis adds a temporal component, considering the energy and resource requirements and impact on the environment for every step that occurs in production and use of a fuel. For example, this might include planting, growing, harvesting, and processing a raw feedstock, then delivery to and use by the consumer, then return of the byproducts to the environment.

BIOFUELS AND SUSTAINABILITY R&D: GAPS AND OPPORTUNITIES FOR INTEGRATION, COORDINATION, AND PARTNERSHIPS

John Carberry (DuPont), Mike Bertolucci (Interface Research Corporation, retired), Emmy Simmons (USAID, retired) and James Fisher (USDA) led the roundtable discussions by suggesting that participants consider the following issues, detailed in Box 1.

Biofuels and Sustainability R&D: Summary of Roundtable Panel Discussions

As each Roundtable group presented the summary of its discussion, several themes emerged. A number of participants emphasized the need for a systems approach to assess biofuels and taking a holistic perspective on the costs and

BOX 1

Gaps and Opportunities for Collaboration in Biofuels R&D Related to Sustainability

- What are the drawbacks of various biofuels, including possible unintended and potentially harmful consequences? What are the potential barriers to expanded use of biofuels as a replacement for gasoline?
- How can socioeconomic and political factors be integrated with our biological and physical knowledge on biofuels? Biofuel production has tended to be viewed as a physical problem of production, efficiency, and delivery. But it's actually much bigger and more complex than that. It's also a business requiring a clear understanding of the market for biofuels. Biofuels are also an international issue, affecting both industrialized and developing countries. The markets overseas may have different pressures and needs. For example, Europe has a higher demand for biodiesel, while domestically the focus is on ethanol.
- How does the demand for biofuel feedstocks affect other markets? For example, the volume of U.S. food aid has dropped by half in the last 5 years, even though the budget has stayed the same? This is a direct result of fuel markets: (1) transport fuels are much more expensive, and (2) grain foods are pricier now because of the increased demand for energy crops.
- How can government, industry, and academia work together so that faster progress can be made?
- What is the potential for GMOs to increase the supply of biofuel feedstocks and what are the technical and public acceptances barriers?
- What key opportunities in biofuels R&D could make the greatest contribution to increasing the availability of alternative transportation fuels? Are there R&D areas that are overlooked or underfunded?

benefits of various energy supply options. They suggested the importance of not only looking at the production of biofuel feedstocks but also at refineries, distribution systems, markets, and consumers. To date, most of the attention has focused on the production side of the equation with little attention to distributions systems—integrating biofuels into larger transportation system, markets, and customers. There is currently no infrastructure in place to make large scale substitutions of biofuels for conventional gasoline. Participants also noted that there are only a limited number of flex fuel vehicles and few outlets selling E-85 or other alternative fuels. Furthermore, there seems to be little understanding of consumer behavior, specifically under what conditions consumers may be willing to buy flex fueled vehicles or to retrofit existing vehicles in order to use higher blends of ethanol and gasoline. A holistic approach also requires considering the whole lifecycle of a fuel. This includes the broad ecosystem perspective discussed above, which considers a feedstock's effects on all of the resources of an ecosystem. A lifecycle analysis adds a temporal component, considering the energy and resource requirement and impact on the environment of every step that occurs in production and use of a fuel. For example, this might include planting, growing, harvesting, and processing a raw feedstock, then delivery to and use by the consumer, then return of the byproducts to the environment. Each potential feedstock and fuel needs to be judged on this basis

A number of participants suggested that it would be useful to develop a framework for assessing biofuels and other alternative fuels in the context of other societal concerns in order to understand the associated risks and benefits and examine the full range of environmental and economic effects. Others emphasized the importance of maintaining a diverse portfolio of fuel sources to reduce the potentially devastating effects of crop failures.

Discussants emphasized the need to take a broad view, looking at a full range of energy choices that are likely to change over time. They stressed the need to maintain flexibility and not get locked into promising but unproven approaches. The development of alternative fuels can be seen as an evolutionary process. It is gradually developing, but we need to identify and remove barriers to facilitate this. Diversified supplies are needed to tackle this large problem. There is no one "silver bullet" that will meet all needs in every region, so we need to pursue a variety of feedstocks.

Many participants stressed the importance of using place based research to assess ecosystem effects as well as to measure direct and indirect costs. The sustainability of biofuel production depends in large measure on the local geography, current climate patterns as well as the potential impact of climate soil productivity, farming techniques, transportation systems, distances from refinery facilities, distances from ultimate customers, costs, and availability of other fuels. In some places bioenergy crops will require large increases in chemical fertilizers and water use with important impacts on local water resources and ecosystems. At the same time there may be impacts outside of the local area. The impacts of increased nutrient loading on the Gulf of Mexico are a prime example with a dramatic expansion of the "dead zone."

Several participants expressed the need for expanded communications between scientists, policy makers, farmers, businesses, and investors. We have individual knowledge that we don't have collectively. Sharing knowledge is critical to keeping expectations in check and meeting long term goals for expanded energy supplies and meeting the transition to sustainability.

Forum participants were also concerned with more fully understanding the political, economic, and social implications of biofuels both domestically and internationally. Who benefits from the current expansion of corn-based biofuel? What changes are likely to occur in the U.S. and internationally? Is production likely to expand into lands already part of the Conservation Reserve Program?

Are the beneficiaries primarily large scale corporate farms or small farms? What is the effect of expanded demand for agricultural land on the viability of small farm? What are the international trade implications of expanding biofuels? Most participants do not view corn-based biofuels as a long term energy solution, therefore, what will be the implications for local producers, rural communities, and investors of a shift to cellulosic ethanol of other alternative fuels?

4

Common Themes

The workshop presentations, panel, and roundtable discussions identified a number of common themes related to sustainability research and development activities on ecosystem services and biofuels. These are summarized below. While they are representative of the views expressed by many of the participants, they do not constitute consensus conclusions of the steering committee.

FRAMING SUSTAINABILITY

Many participants strongly cautioned against devoting too much effort to refining the definition of sustainability, noting that it was more important to emphasize the relationship between people and life support systems. Workshop participants also suggested that it might be easier to recognize that certain actions and trends are *not* sustainable over the long term rather than to describe those that are. In other words, moving away from unsustainable practices is a way to make the transition toward sustainability. Regardless of the topic, a number of participants stressed the importance of maintaining a sustainability perspective when determining research priorities and developing policies and programs.

REFOCUSING STRATEGIES

Many agencies are now using a sustainability lens to help focus their R&D activities as well as their natural resource management programs. For example, the USGS science strategy highlights plans for work on ecosystem services and biofuels. The EPA has issued a *Sustainability Research Strategy*[1] (1) to improve understanding of the earth's natural and manmade systems, (2) to assess threats to these systems, (3) to design and apply cost effective industrial processes and (4) to develop and apply new technologies and decision support tools. The EPA is also beginning the development of a strategy for sustainable biofuels. The Forest Service is now using the concept of ecosystem services as a framework for describing the benefits of forests, for evaluating the effects of policy and management decisions and for advocating the use of economic incentives to protect private forest lands from development. Both the USDA and the National Science Foundation currently have sustainability councils to help guide R&D decisions. The USDA Council on Sustainable Development includes representatives from all the mission agencies and is focused on policies and programs supporting sustainable agriculture, sustainable forestry, and sustainable rural communities.

KEY QUESTIONS

While many of the workshop discussions focused on research gaps, participants emphasized that there is much we already know especially regarding the natural sciences associated with ecosystem services and biofuels. The most significant gaps are in understanding the associated social, economic, political, and behavioral issues. Some specific research questions are listed below:

- Is it possible to more clearly identify the effects of changes in ecosystem conditions on communities and vulnerable people?
- What are the implications of expanded U.S. ethanol production for changes in habitat and biodiversity?
- What are the economic and social impacts of biofuel production on rural communities and states?
- Are available indicators matched with needs of local resource managers? Many participants noted the importance of developing indicators and metrics to evaluate programs, to track changes in ecosystem conditions, to assess pressures and drivers, to warn of potential vulnerabilities and "tipping points."
- How can the concept of ecosystem services be made "real"? Participants acknowledged that efforts to value ecosystem services were helpful but suggested that better information was needed to educate stakeholders about the functions of ecosystem processes and services and the benefits and costs associated with human interactions.

[1]EPA's *Sustainability Research Strategy*, Foreword, Accessed on 3/19/08, http://www.epa.gov/sustainability/pdfs/EPA-12057_SRS_R4-1.pdf

HOLISTIC APPROACH

For both ecosystem services and biofuels, many participants emphasized the need to maintain a big picture or holistic perspective—drawing on multiple disciplines, focusing on different geographic and temporal scales as well as recognizing the needs of a diverse set of stakeholders. For example, in the case of biofuels, some participants suggested that biomass R&D efforts should be considered as one part of a mix of energy supplies or an energy portfolio. Furthermore, bioenergy R&D efforts should focus on more than the development and conversion of feedstocks to understand the effects of production and use on critical natural resources—water (quality and quantity), soils, and direct/indirect land use—as well as the impacts on transportation, local communities, and vulnerable populations. In addition, many participants noted important trade-offs between fuel feedstocks and food (for people as well as animals) both domestically and internationally. There was considerable emphasis on the value of place-based research related to both biofuels and ecosystem services as well as the link between increased production of biofuels and the vitality of local ecosystem services.

INCREASED INTERAGENCY COLLABORATION

There is considerable R&D collaboration among agencies in addressing both ecosystems services and biofuels. However there are additional opportunities for collaborative activities and to leverage activities of other agencies, especially at a local or regional level. The Committee on Environment and Natural Resources (CENR), Ecosystem Services Task Team of the National Science and Technology Council[2] (NSTC) provides one mechanism to share information on research related to ecosystem management activities and to identify opportunities for cooperation and collaboration. However, some participants noted that the CENR needs to be revitalized to take full advantage of the programs and new staff at the Forest Service, EPA, USGS, and other agencies.

Several workshop participants also suggested that the *Ocean Science Plan* is a possible model for setting federal research priorities on ecosystem services, supporting federal funding requests and strengthening collaboration among agencies. However, they noted the difficulty in uniting agency budgeting processes.

In the case of biofuels, R&D priorities are coordinated by an interagency board. The Biomass R&D Board, co-chaired by DOE and USDA, includes cabinet level representatives from DOI, DOT, EPA and the Commerce Department. To date, R&D efforts have focused largely on feedstocks and conversion technologies. Sustainability issues have not been a major focus. However, John Mizroch, the DOE Deputy Assistant Administrator for Energy Efficiency and Renewable

[2]The NSTC is led by the President's Science Advisor and is responsible for coordinating the federal S&T policy making process; the ensuring the S&T policy decisions and programs are consistent with the President's goals, and integrating the President's S&T agenda across agencies.

Energy, announced that the "Billion Ton" study was to be updated with more attention to sustainability.[3] In addition, a number of agencies represented at the Forum have significant research efforts underway to understand some of the key sustainability issues associated with the production and use of biofuels.

KNOWLEDGE DISSEMINATION

Participants from the academic and industrial communities noted the value of making information/knowledge from federal R&D activities more widely available. They explained that results of federal research activities are often not published in scientific journals, and thus dissemination is rather limited. A few participants suggested that the *Proceedings of the National Academy of Sciences'* section on sustainability science[4] might be used to facilitate more widespread dissemination. There was also talk about the need to scale-up knowledge from local R&D activities for application at state and national levels.

LINKING KNOWLEDGE WITH ACTION

Participants emphasized the need to focus on linking existing scientific and technical information to programs and policies at federal, regional, and local levels. While the knowledge-action link is often cited as a potential barrier in making the sustainability transition, it may be somewhat less of a barrier for federal researchers. Agencies such as USDA and DOI are directly responsible for managing federal lands (USDA, including the Forest Service, and DOI manage some 55 percent of US lands). The EPA uses the results of their R&D activities to support their regulatory programs, creating natural links to customers and constituents.

A number of barriers, however, remain in linking knowledge to program and policy actions. Constraints include:

- Limited understanding of how decisions are made at the local/regional level and of the importance of politics and institutions at all levels. For example, political incumbents may be reluctant to shift policies or promote new, more sustainable programs if these might undermine their electability. There are also problems in integrating across political scales (local to federal) as well as across local jurisdictions. Many participants noted the importance of connecting scientists and local stakeholders, possibly through bridging institutions or networks, and of recognizing

[3]Biomass as Feedstock for a Bioenergy and Bioproducts Industry: The Technical Feasibility of a Billion Ton Annual Supply, Oak Ridge National Laboratory for the US Department of Energy and the Department of Agriculture, April 2005.

[4]PNAS, http://www.pnas.org/misc/sustainability.shtml. Accessed on: 1/7/2008

incentives and disincentives for working together (who benefits) and understanding how to promote trust.

- A number of participants talked about uncertainty and/or the lack of complete information as a barrier to action. They suggested that it was important to take some initial steps, setting the stage for future action with what we already know works, while continuing to support R&D. Many environmental changes are happening with potential long term negative consequences. Programs and policy choices offer win-win opportunities as shown by approaches to addressing global climate change such as improvements in energy efficiency.

NEXT STEPS

Many participants suggested a number of steps that they might take with their own organizations or that could be undertaken by the National Academies. Some of the ideas discussed included:

- Convening a regional forum with federal, state, and private partners to explore how R&D efforts are used (or not) to determine policies and local management programs, to identify the barriers to more effective local ecosystem services management efforts—including political, economic, and social barriers—and to understand the research needed by local decision makers.
- Developing a set of indicators related to ecosystem services—something like a dashboard—that would reflect the status of critical ecosystems. These indicators could then be used to identify specific management priorities and provide a basis for examining political, economic and environmental trade-offs of various management strategies.
- Exploring the possibility of creating regional centers of excellence to pool limited agency R&D resources and to create linkages between the research community and local decision makers.
- Reinvigorating the NSTC Committee on Environment and Natural Resources, specifically the CENR Ecosystem Services Working Group to share information about R&D being done by different agencies, encourage collaborative activities and communicate best practices.
- Creating a framework for assessing bioenergy production and biorefineries in the context of sustainability. The emphasis would be on identifying key issues along the supply chain, developing metrics, and focusing on key research and development opportunities.
- Expanding place-based studies on biofuel production and use, recognizing that soils, climate, water availability, and feedstock choices will have unique economic, social, and environmental impacts.
- Examining changes required in the U.S. transportation infrastructure

to accomodate increased biofuel production, and the socio-economic, environmental, safety and health consequences associated with such changes and identifying needed R&D.

- Convening a series of workshops between the science community and the investment community. This would provide an opportunity for the investment community to be exposed to the scientific perspectives which were a highlight of the recent Federal Forum and help inform investment decisions that are likely to have a significant impact on prospects for long term sustainability. One topic could be biofuels with a focus on cellulosic ethanol providing an overview of the state of the science and technology, risks, and trade-offs.
- Hosting a second federal R&D Forum looking at R&D activities related to oceans and coastal areas, possibly including the implementation of the new Ocean Research Priorities Plan, management issues in the Arctic, and examining the priorities of other countries for managing the oceans.

Appendixes

A

Transitioning to Sustainability through Research and Development on Ecosystem Services and Biofuels

The National Academies' First Federal Sustainability Research and Development Forum
October 17-18, 2007

Location: The National Academy of Sciences (Members Room)
2100 C Street NW
Washington, DC

October 17, 2007

SESSION ONE:
INTRODUCTION AND OVERVIEW (1 ½ HOURS)

8:30 am **Opening Remarks, Introductions, and Goals of the Forum** (Pamela Matson, Co-Chair, Roundtable on Science and Technology for Sustainability)

9:00 am **Framework for Sustainability Research and Development** (Pamela Matson)
General characteristics of research and development to support the transition to sustainability.

9:30 am **Discussion**

SESSION TWO:
ECOSYSTEMS SERVICES AND SUSTAINABILITY (4 HOURS)

9:45 am **Ecosystem Services: Challenges and Opportunities to Sustainability** (Steve Carpenter, University of Wisconsin, Madison)

10:15 am **BREAK**

10:30 am **Federal Policies and Research Priorities Related to Ecosystem Services** Overview from CENR working group on ecosystem services. (Bruce Rodan, U.S. Office of Science and Technology Policy and Iris Goodman, Co-Chair, CENR Ecosystems Services Working Group)

10:50 am **Ecosystem Services—State-of-the-Art Examples of Sustainability Research and Development** (Panel leader: Kai N. Lee, Packard Foundation)
Panelists:
- Ned Euliss, US Geological Survey
- Mark Nechodom, US Forest Service/Pacific Southwest Research Station
- Iris Goodman, US Environmental Protection Agency
- Dan Kugler, US Department of Agriculture
- Steve Murawski, National Ocean and Atmospheric Administration
- Woody Turner, National Aeronautics and Space Administration
- Margaret Palmer, University of Maryland Center for Environmental Sciences

12:00 pm **Working Lunch—**Informal Discussions on State of the Art Ecosystem Services' Examples

12:45 pm **Ecosystem Services R&D Gaps, Opportunities for Integration, Coordination, and Partnerships** (Panel leader: William C. Clark, Harvard)
Panelists will reflect on gaps and opportunities based on their review of meeting materials, previous presentations, and personal experience. Panel leader will tee-up questions for round table discussions on gaps and opportunities
Panelists:
- Ann Bartuska, U.S. Forest Service
- Sara Scherr, Ecoagriculture Partners

1:15 pm **Research and Development Gaps and Opportunities—Roundtable Discussions** (8-10 persons per table)
Participants will discuss questions about research and development opportunities and gaps related to ecosystem services. Examples of questions:
- Are there major research and development gaps? How well do agency research and development programs address the key

issues related to sustainability and ecosystem services (e.g., is anyone working on resilience or vulnerability in the provision of ecosystem services? How well is the knowledge-action link included in research endeavors?)

- Are there areas of significant overlap or gaps among federal agencies and other organizations, and if so, what strategies could be used to foster effective collaboration in these areas?
- Are new analytical tools or data needed?

2:30 pm **Feedback from Roundtable Discussions to Entire Group and Preliminary Synthesis** (Discussion leader: Bill Clark)

3:15 pm **BREAK**

SESSION THREE: BIOFUELS AND SUSTAINABILITY (4 HOURS— TO BE CONTINUED ON DAY 2)

3:30 pm **Biofuels: Challenges and Opportunities to Sustainability** (Daniel M. Kammen, University of California, Berkeley)

4:00 pm **Federal Policies and Research Priorities Related to Biofuels** (John Mizroch, Principal Deputy Assistant Secretary, Office of Energy Efficiency and Renewable Energy, DOE)

4:20 pm **Biofuels—State-of-the-Art Examples of Sustainability Research and Development** (Panel leader: Todd Mitchell, STS Roundtable Member)

Panelists:

- Jeff Steiner, US Department of Agriculture
- Richard B. Alexander, US Geological Survey
- Randy Bruins, US Environmental Protection Agency
- Marcia Patton-Mallory, US Forest Service
- William Chernicoff, US Department of Transportation
- Jacques Beaudry-Losique, US Department of Energy

5:30 pm **Summary and Plan for the Next Day**

5:40 pm **Adjourn for Day**

Informal Reception

October 18, 2007

8:30 am **Welcome and Brief Recap of Day 1** (Pamela Matson)

SESSION THREE: BIOFUELS AND SUSTAINABILITY (CONTINUED)

8:45 am **Water Implications of Biofuels (Report Briefing)** (Steve Parker, The National Academies)

8:55 am **Biofuels R&D Gaps, Opportunities for Integration, Coordination, and Partnerships** (Panel leader: John Carberry, E. I. du Pont de Nemours & Company)
Panelists will reflect on gaps and opportunities based on their review of meeting materials, previous presentations, and personal experience. Panel leader will tee-up questions for round table discussions on gaps and opportunitics
Panelists
- Mike Bertolucci, Interface Research Corporation
- Emmy Simmons, U.S. Agency for International Development (retired)
- James Fischer, U.S. Department of Agriculture

9:15 am **Research and Development Gaps and Opportunities—Roundtable Discussions** (8-10 persons per table)
Participants will discuss questions about research and development opportunities and gaps related to biofuels. Examples of questions:
- Are there major research and development gaps? How well do agency programs address the key challenges to the sustainable production and use of biofuels? How well is the knowledge-action link included in research endeavors?
- Are there areas of significant overlap or gaps among federal agencies and other organizations, and if so, what strategies could be used to foster effective collaboration and coordination in these areas?
- Are new analytical tools or data needed?

10:30 am **BREAK**

11:00 am **Feedback from Roundtable Discussions to Entire Group and Preliminary Synthesis** (Discussion Leader: John Carberry)

12:00 pm **Working Lunch Presentation—Ocean Research Priorities Plan and its Implications for Sustainability** (Dan Walker, Senior Policy Analyst, U.S. Office of Science and Technology Policy)

SESSION FOUR: SUMMARY AND FORUM WRAP-UP (2 HOURS)

1:00 pm **Lessons Learned beyond the Biofuels and Ecosystems Services Topics**
Panelists will summarize major gaps and collaboration opportunities from the discussions of ecosystem services and biofuels, as well as the major commonalities and differences across the two topics.

2:30 pm **Closing Remarks**

B

List of Organizers and Panelists

Richard B. Alexander
Hydrologist
U.S. Geological Survey

Jeffery Steiner
National Program Leader
U.S. Department of Agriculture, ARS

Ann Bartuska
Deputy Chief for Research & Development U.S. Forest Service

Diana Bauer
Environmental Engineer
U.S. Department of Transportation

Jacques Beaudry-Losique
Program Manager
U.S. Department of Energy

Mike Bertolucci
President
Interface Research Corporation

Randy Bruins
Office of Research and Development
U.S. Environmental Protection Agency

John Carberry
Director of Environmental Technology
E. I. du Pont de Nemours & Company

Steve Carpenter
Professor
University of Wisconsin, Madison

William Chernicoff
Engineer
U.S. Department of Transportation

William Clark
Harvey Brooks Professor of International Science
Kennedy School of Government
Harvard University

Gregory Crosby
National Program Leader
Natural Resources and Environment Unit
U.S. Department of Agriculture (CSREES)

Ned Euliss
Wildlife Biologist
U.S. Geological Survey

James Fischer
Senior Scientific Advisor
REE Directorate
U.S. Department of Agriculture

Iris Goodman
Environmental Scientist
U.S. Environmental Protection Agency

Linda Gundersen
Chief Scientist for Geology
U.S. Geological Survey

Alan Hecht
Director of Sustainable Development
Office of Research and Development
U.S. Environmental Protection Agency

Daniel Kammen
Professor
University of California, Berkeley

Dan Kugler
Deputy Administrator
Natural Resources and Environment
U.S. Department of Agriculture

Kai Lee
Program Officer
Conservation & Science Program
The David and Lucille Packard Foundation

James Mahoney
Former Assistant Secretary of Commerce for Oceans and Atmosphere/Deputy Administrator
National Oceanic and Atmospheric Administration

Pamela Matson
Dean of the School of Earth Sciences
Stanford University

John Mizroch
Principal Deputy Assistant Secretary
Office of Energy Efficiency and Renewable Energy
U.S. Department of Energy

Steve Murawski
Director of Scientific Programs and Chief Science Advisor for NOAA Fisheries
National Ocean and Atmospheric Administration

Mark Nechodom
Pacific Southwest Research Station
U.S. Forest Service

Margaret Palmer
Professor
University of Maryland Center for Environmental Sciences

Steve Parker
Director
Water, Science and Technology Board
The National Academies

Marcia Patton-Mallory
Biomass and Bioenergy Coordinator
U.S. Forest Service

Bruce Rodan
Senior Policy Analyst
U.S. Office of Science and Technology Policy

Sara Scherr
President, CEO
Ecoagriculture Partners

Emmy Simmons
Former Assistant Administrator for Economic Growth, Agriculture, and Trade
U.S. Agency for International Development

Woody Turner
Program Scientist
National Aeronautics and Space Administration

Dan Walker
Senior Policy Analyst
U.S. Office of Science and Technology Policy

Staff

Patricia Koshel
Senior Program Officer
Science and Technology for Sustainability
The National Academies

Kathleen McAllister
Senior Program Assistant
Science and Technology for Sustainability
The National Academies

Marty Perreault
Director
Roundtable on Science and Technology for Sustainability
The National Academies

Gregory Symmes
Deputy Executive Director
Division on Earth and Life Studies
The National Academies

Derek Vollmer
Senior Program Associate
Science and Technology for Sustainability
The National Academies

Julia Kregenow
Christine Mirzayan Science and Technology Policy Graduate Fellow
The National Academies

C

Descriptions of Agency Activities Presented at the Forum on Ecosystem Services and Sustainability[1]

[1]Presentations are available online at http://sustainability.nationalacademies.org/Forum.shtml

TITLE OF PROJECT OR PROGRAM:
Ecosystem Services in the Prairie Pothole Region: Impacts of Management and Climate Change

AGENCY:
U.S. Geological Survey

PROJECT/PROGRAM DESCRIPTION:
The Prairie Pothole Region of the United States and Canada is a unique area where shallow depressions created during Pleistocene glaciation interact with mid-continental climate variations to create a variety of wetlands that supply a suite of ecosystem services. These unique wetlands comprise a diversity of aquatic invertebrates and vertebrate wildlife that depend upon them as food. The seasonal wetlands serve as safe breeding grounds for a significant population of ducks and an important stopover for migrating shorebirds. In the past century large portions of the area have been transformed to cropland. What remains of the glaciated wetlands supports more than 300 bird species, producing about half of North America's 40 million ducks.

USGS Northern Prairie Wildlife Research Center and USDA's Agricultural Research Service have also collaborated on a long term study to understand the potential of prairie pothole region wetlands to sequester carbon emitted into the atmosphere from the burning of fossil fuels. Results suggest that wetlands traditionally functioned as sinks for atmospheric carbon, but cultivation has shifted their function to be sources of atmospheric carbon. Data suggest that equal or greater amounts of atmospheric carbon can be stored in wetlands through restoration programs when compared with cropland, even though the acreage of wetlands is much smaller. Further, nitrous oxide emissions are reduced for every acre going back to wetland because of the reduction in fertilizer use.

Finally, a new project has been initiated because of the unique resources at the USGS and the ability to integrate ecosystem-based research across multiple disciplines and large areas. The unique resources include:

1) biogeochemical modeling which emphasizes agricultural practices and simulates sustainability and impacts on ecosystem goods and services,
2) socioeconomic modeling of land use and land use trends across broad regions,
3) dynamic monitoring of Ecosystem Performance and the Net Ecosystem Exchange of carbon,
4) access to archival remote sensing data and near-real time data from a variety of sources,
5) capability to provide large datasets to the user community in a seamless manner, and

The goal of this research is to evaluate the effects of an expanded agricultural program for biofuels and concurrent changes in climate on ecosystem sustainability across the Northern Great Plains. We will develop credible land-change scenarios to project alternative landscape futures through 2050 and will analyze the results to estimate effects on ecosystem processes and services. "Ecosystem services" frequently denotes services to humans. We are using the term to encompass services realized by any component of the ecosystem, such as by wildlife. This research will enable us to address the following key questions:

1. How might landscape patterns change in response to demand for expanded biomass production?
2. What are the environmental consequences (on biogeochemical cycling, soil erosion, nutrient transport to waterbodies, greenhouse gas emissions, quantity and quality of wildlife habitat) of biomass production for energy?
3. What are the full costs and benefits of biomass production for energy, including agricultural sector profitability?
4. How will projected climate change impact agricultural production and profitability?
5. How will projected climate change impact the provision of ecosystem services, both directly and indirectly through changes in landscape patterns?
6. What are the feedbacks among land-use change, economic and policy drivers, climate, biophysical processes, and a variety of ecosystem services?
7. What is the Net Ecosystem Exchange of carbon and energy associated with each potential land use for biomass and what is the 'end to end' total energy/carbon balance?
8. What are the special concerns about habitat quality and wildlife concerns in this region?
9. What are the most important factors and constraints in implementing a long-term sustainable biomass-for-energy industry?

Our landscape scenarios will highlight four crop types under research for biofuel: corn, soybeans, switchgrass, and mixed prairie grasses. We will project landscape change under current climate, low climate change, and high climate change, as predicted by output from major global climate models. Results from the landscape scenarios will be assessed relative to ecosystem quality, processes, and services. We will use the assessments to determine the balance of economic performance (net value of biofuel and agricultural production) with ecosystem sustainability.

The Integrated Landscape Monitoring—Prairie Pilot is one of four science thrusts initiated by the U.S. Geological Survey in 2006. The goal of each pilot is to develop a regional monitoring framework to model and monitor the performance of conservation programs, especially their provision of specific ecosystem goods and services (e.g., carbon sequestration, greenhouse gas flux, flood water storage, water quality, erosion reduction, wildlife habitat). The Prairie Pilot is unique in that it was initially developed to include the specific needs of diverse land management agencies in the Departments of Agriculture and Interior. An inter-agency science team (Farm Services Agency, Natural Resources Conservation Service, and the U.S. Fish and Wildlife Service) was established to define a list of specific ecosystem goods each agency wished to use as measures of performance of conservation programs. Those ecosystem goods and services will be modeled collectively to incorporate the delivery of multiple and simultaneous outcomes of conservation programs to provide the comprehensive view required to predict unintended and potentially negative, consequences of land-use change. To accurately evaluate program performance, the model will be designed to separate change in ecosystem services due to natural factors (e.g., dynamic mid-continental climate) from those attributable to federal conservation programs. The basic modeling framework is based on the unique climatic drivers in the Prairie Pothole ecosystem and it will provide a transparent means of incorporating the best available scientific information into a decision support tool to facilitate consistent evaluations and forecasts of program performance by different agencies and other users.

The following diagram is presented to illustrate our integrated approach. Our overall societal goal is to define the most appropriate and sustainable land uses that maintain appropriate ecosystem goods and services and to convey this information to managers and policy makers as we provide model outputs for alternative landscape futures.

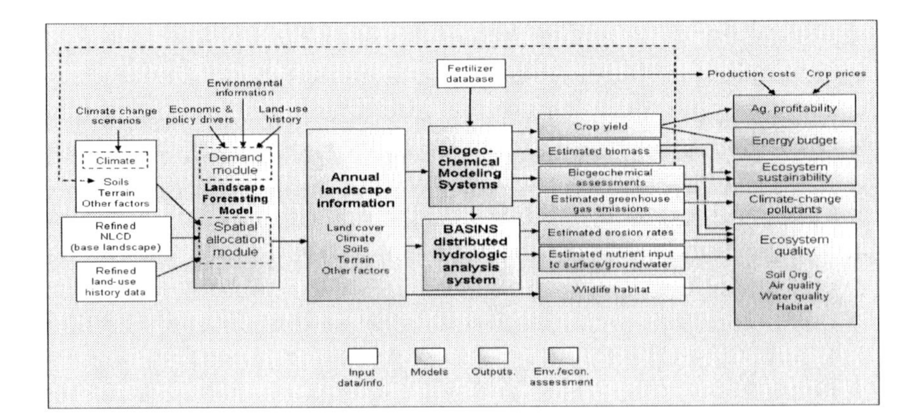

DESCRIBE ANY NOTABLE RESULTS, OUTCOMES OR IMPACTS TO DATE, IF ANY:

Built upon extensive previous work (http://www.npwrc.usgs.gov/about/factsheet/wetlands.htm) on agricultural practices, ecosystem services response to climate change and carbon retention in the Prairie Potholes Region, the new project will start in FY08.

PERFORMERS/OTHER PARTNERS (FEDERAL, STATES, OR LOCAL):

USGS, Bureau of Land Management, USDA, SunGrant program of DOT, EPA, and DOE Regional Program (PCOR) and university research scientists, U.S. Fish and Wildlife Service, Farm Services Agency, Natural Resources Conservation Service.

PROJECT PERIOD:

Over 10 years of previous work, new project starts in 2008.

TITLE OF PROJECT OR PROGRAM:
Alder Springs Fuels Reduction Stewardship Program

AGENCY:
USDA Forest Service, Pacific Southwest Research Station

PROJECT/PROGRAM DESCRIPTION:
In 2006, the Mendocino National Forest, Pacific Southwest Research Station (PSW), and Winrock International Institute for Agricultural Development received funding from the Forest Service to partner with the State of California for a project that will demonstrate and evaluate potential market opportunities for carbon sequestration and carbon offsets. This project will take place in conjunction with the Alder Springs Fuels Reduction Stewardship Project and will monitor fuels management treatments in order to accomplish the following:

- Quantify greenhouse gas emission reductions resulting from fuels management treatments
- Evaluate and quantify potential revenues in current and future carbon markets
- Evaluate potential for renewable energy credits and incentives associated with biomass energy production

This project is of great relevance due to increasing interest in carbon management, renewable energy production, and because of the magnitude of National Forest System lands in need of fuels reduction treatments. The Forest Service is supporting this project in order to assess the potential for generating possible market incentives for fuels and forest health treatments, which would help extend landscape treatment capabilities.

This fuels project and the associated research are important first steps toward understanding how public forest management might contribute to mitigating global climate change. Although research models already suggest that there are likely climate change benefits to be gained from forest management, this is the first time those models are being tested on an actual forest management project.

There are three ways in which a fuels reduction project might reduce greenhouse gas emissions:

1) Thinning the forest improves forest health, and a healthy forest absorbs more CO_2 from the atmosphere,
2) Thinned forests are less likely to experience catastrophic wildfires that release vast amounts of greenhouse gasses into the atmosphere,
3) When the biomass from a thinning project is used to generate electricity, the net amount of carbon released into the atmosphere is considered

"carbon neutral" when compared to fossil fuels that are used to generate the same amount of energy.

The Alder Springs project was developed to create landscape change throughout areas considered at high risk for catastrophic wildfire for both wildland urban interface protection and ecosystem health. Although the carbon research associated with this project is important, this project will also accomplish essential hazardous fuels reduction work for the Mendocino National Forest. Stewardship contract authorities permit the Forest Service to trade goods for services; that is, it allows private organizations or businesses to remove forest products such as trees, undergrowth and biomass in return for performing work to restore and maintain healthy forest ecosystems.

Carbon markets coupled with State and Federal incentives could produce significant market opportunities for the private sector associated with fuels reduction projects. These opportunities would allow forest managers to extend programs and treat more acres, improving forest health and reducing the threat of wildfire. The Forest Service is supporting this project in order to assess the potential for generating market incentives for fuels treatments, which would help extend our landscape treatment capability.

DESCRIBE ANY NOTABLE RESULTS, OUTCOMES OR IMPACTS TO DATE, IF ANY:

None yet

PERFORMERS/OTHER PARTNERS (FEDERAL, STATES, OR LOCAL):

Mendocino National Forest, Pacific Southwest Research Station, California Department of Forestry and Fire Protection, California Energy Commission, Winrock International, Wheelabrator Shasta Energy, Inc., Future Resources Associates, TSS Consultants, and Sierra Pacific Industries

PROJECT PERIOD:

Research work period: Spring 2007 – March 2009
Project work period: July 2007 – March 2009
Research finding to be published 2 to 3 years following project completion

PARTNER FUNDING LEVELS (CURRENT OR PROPOSED):

$250,000 — Winrock International
$50,000 — Future Resources Associates (Dr. Gregg Morris for carbon modeling)
$50,000 — TSS Consultants (Fire modeling and biomass power marketing consulting)

TITLE OF PROJECT OR PROGRAM:
Ecosystem Services Research in Communities: Willamette River Basin Study

AGENCY:
U.S. Environmental Protection Agency

PROJECT/PROGRAM DESCRIPTION:
EPA's Ecological Research Program (ERP) in the Office of Research and Development (ORD) is focused on the study of ecosystem services and the benefits to human well-being provided by ecological systems. As part of the ERP's community-based research, this project will identify and characterize ecosystem services in the Willamette River Basin, located in Oregon between the Cascade mountains and the Pacific Ocean. This basin is primarily in forest and agriculture (forests/forestry comprises about 56%; agriculture is about 20% of land cover). The Willamette Basin's population is expected to double by 2050. There is considerable local interest in sustainable economic growth and resource utilization.

The research goal for the Willamette River Basin is to quantify the area's ecosystem services and understand the effects of man-made stressors on those services. Understanding these interactions will help local decision makers understand the ecological costs and benefits of existing and proposed land management and growth policies. The study will focus on major ecosystem service pertinent to land cover categories of agriculture, forests and riparian wetlands.

The goals of the initiative are to:

- Identify critical knowledge gaps in the ecological processes underlying ecosystem services
- Map ecosystem services in the river basin based on current conditions and available data
- Quantify the response of ecosystem services to current and projected conditions and stressors (i.e., land use changes, climate change, cropping practices, etc.)
- Quantify linkages and trade-offs among bundles of ecosystem services in response to land use, climate, and other variables
- Model the future responses of ecosystem services to probable future conditions
- Determine how these changes in ecosystem services affect human well being.

DESCRIBE ANY NOTABLE RESULTS, OUTCOMES OR IMPACTS TO DATE:

Previous research in the Willamette River Basin conducted by EPA's ERP developed alternative future scenarios and estimated how alternative development practices would likely affect a number of ecological endpoints. These results were portrayed as maps of conservation and restoration opportunities and were used by the Willamette Restoration Initiative in targeting various large scale restoration and "services trading" initiatives. The present study builds on this foundation and will expand and refine ecosystem services maps, models of ecological production functions, and creation of decision support tools, with a particular focus on riparian systems.

Research clients include U.S. EPA Region 10 office in Seattle, Washington, which has regulatory authority in the Willamette River Basin, the Oregon Department of Environmental Quality, and local municipalities. Using these tools, decision makers can implement proactive policy and management decisions over time and at multiple scales. The research also will be integrated with other ERP community-based ecosystems research to create a suite of methods and tools for evaluating ecosystem services that can be transferred to other EPA regions and national program offices.

PERFORMERS/OTHER PARTNERS (FEDERAL, STATES, OR LOCAL):

NA

PROJECT PERIOD:

Start Date: 2005 End Date: 2014

FUNDING LEVELS (CURRENT OR PROPOSED):

This research is being carried out primarily by EPA ERP's in-house scientists; a portion of this research is being conducted by scientists at Oregon State University and University of Oregon, under the sponsorship by ERP's extramural STAR grant program.

TITLE OF PROJECT OR PROGRAM:
Long-term Agro-ecosystem Research

AGENCY:
USDA/Cooperative State Research, Education, and Extension Service (CSREES)

PROJECT/PROGRAM DESCRIPTION:
The long-term goal of the program is to support inter-disciplinary experimental, observational, theoretical, and modeling studies that can create improved cropping and tillage systems that supply high quality food, fiber and fuel while reducing agriculture's impact on the environment. Currently, one of the limitations in achieving this goal is a lack of understanding concerning long-term processes and the coupled dynamics of ecological, production, and socio-economic systems.

The focus of the program will be on soil carbon management. Soil is the largest reservoir of carbon in terrestrial ecosystems. Understanding the mechanisms and processes involved in the accumulation and loss of stored soil carbon provides an opportunity to develop management strategies that increase carbon storage and decrease carbon loss. Soil carbon is relevant to food security, economic viability of farms, and climate change.

Key issues to be addressed include how the management of agronomic inputs impacts soil carbon storage; how the maximum potential carbon storage of a soil can be estimated; how long it takes to attain the storage potential; how long it resides; what the regional differences are; how changes in the global environment, such as increased atmospheric carbon dioxide levels and weather patterns, impact soil carbon cycling; the role of the carbon-to-nitrogen ratio of crop residue in greenhouse gas emissions; and the social and economic benefits associated with particular carbon management strategies.

The general objective is the transfer of scientific results to local understanding, acceptance and support based on social, economic, and environmental benefits of sequestering carbon. Emphasis will be placed establishing a community network of farmers, researchers and extension personnel and measuring soil carbon storage and understanding carbon dynamics of different farming practices.

DESCRIBE ANY NOTABLE RESULTS, OUTCOMES OR IMPACTS TO DATE, IF ANY:
N/A

PERFORMERS/OTHER PARTNERS (FEDERAL, STATES, OR LOCAL):
National Science Foundation

PROJECT PERIOD:
Should begin in 2008

FUNDING LEVELS (CURRENT OR PROPOSED):
Proposed $1 million/year.

TITLE OF PROJECT OR PROGRAM:
Implementing a holistic "Ecosystem Approach" to NOAA's Coastal and Marine Mandates—Providing the Science Base to Achieve Ecosystem Objectives

AGENCY:
National Oceanic and Atmospheric Administration (NOAA)

PROJECT/PROGRAM DESCRIPTION:
NOAA has a wide range of stewardship responsibilities related to ocean, coastal, and Great Lakes natural resource management. In the past these mandates were pursued using science and governance mechanisms that were species or issue-based: a "Single Sector" approach. More recently, the Agency and its external stakeholder groups have advocated a more holistic approach to regional ecosystem governance and science that provides a more comprehensive view of marine ecosystem management: an "Ecosystem Approach."

Advantages of Ecosystem Approach over Single Sector Management:

- Provides a "big picture" of an ecosystem.
- Broad perspective and multiple time and space scales.
- Long-term strategic balance among competing uses of ecosystems.& tradeoffs.
- Human impacts and effects on communities considered in tradeoff analyses.
- Supports Adaptive and integrated management across sectors.
- Shared and standardized observations.

There is considerable interest in Congress to implement more comprehensive management, but responsibilities are currently spread over many different federal agencies and bureaus within agencies, making this integration difficult. Notwithstanding these organizational difficulties, a number of noteworthy external drivers have advocated for ecosystem-based management, and there are numerous proposals in Congress to define the issues, authorities, and processes to implement ecosystem approaches. Chief among these external drivers is the recent reports of the U.S. Ocean Commission, and the private Pew Oceans Commissions reports which both simultaneously have called for the federal government, in concert with the states, local government entities, and NGOs to collaborate on ecosystem management.

What are the elements of an ecosystem approach and how do they differ from business as usual?
NOAA has been a leader in the effort to envision and implement marine EAM.

Briefly, most proponents identify a series of principles inherent in the concept, including the following:

Characteristics of Ecosystem Approaches to Management:

- Adaptive
- Collaborative
- Incremental
- Geographically specific
- Accounts for ecosystem knowledge and uncertainty
- Considers multiple external factors
- Strives to balance diverse societal objectives

While there presently is no overall federal governmental mandate to employ EAM across the Federal ocean agencies, clearly many of these steps can be imbedded into present mandates and working relationships among governmental and extra-governmental entities. There are many such efforts currently ongoing.

In order to accomplish the goals of biodiversity protection, enhancing ecosystem resilience to perturbations and promoting sustainability, a broad set of monitoring, research and forecasting tools are required. In order to help frame management decision making, quantitative "decision support tools" are required to evaluate ecosystem outcomes resulting from alternative options. One such approach to decision support tools is to conduct Integrated Ecosystem Assessments (or IEAs).

What are IEAs?

An Integrated Ecosystem Assessment is defined as "a formal synthesis and quantitative analysis of information on relevant physical, chemical, ecological and human factors in relation to specified ecosystem management objectives." It brings together citizens, industry representatives, scientists, and policy makers through formal processes to evaluate a range of policy and/or management actions on difficult environmental problems. An IEA provides an assessment of baseline conditions and identifies important stressors to the system. It also delivers ecological forecasts and scenario developments under changing ecosystem conditions as well as different management actions. IEAs are an emerging concept under development in the USA, and elsewhere in the world. While our concept shares many attributes with related efforts, NOAA's IEA concept, if implemented as outlined here, will be more comprehensive, complete and useful over a broader constituency than any previous efforts.

The primary objectives of the IEA are to:

- Identify key management or policy questions
- Assess status and trends of the ecosystem

- Assess the environmental, social, and economic causes and consequences of these trends
- Forecast ecosystem responses to climate change
- Forecast likely ecosystem status under a range of policy and/or management actions
- Identify crucial gaps the knowledge of the ecosystem that will guide future research and data acquisition efforts.

An IEA uses approaches that determine the probability that ecological or socio-economic properties of systems will move beyond acceptable limits as defined by management objectives. A useful IEA must provide an efficient, transparent means of summarizing the status of ecosystem components, screening and prioritizing potential risks, and evaluating alternative management strategies against a backdrop of environmental (e.g., climatic, oceanographic, seasonal) variability. An IEA provides a means of evaluating tradeoffs in management strategies among potentially competing ecosystem use sectors.

What is the process for conducting IEAs?

The Drivers, Pressures, States, Impacts, Response (DPSIR) framework is a well accepted model for environmental management supporting a wide variety of disciplines. This can be summarized as: Driver → Pressure → State → Impacts → Response → Driver. . . . The DPSIR framework illustrates the process of IEA development in relation to stated ecosystem problems and goals, and is a model for continuous process improvement supporting an adaptive approach to ecosystem-based management.

The process by which a regional or local entity produces IEAs can be applied to this model is as follows:

1. Identify major human and natural factors affecting ecosystem. Define scale. (Driver/Pressure)
2. Organize relevant date. Select key indicators of ecosystem status. (State)
3. Link ecosystem status indicators to drivers and pressures using ecosystem models. (State/Impacts)
4. Evaluate ecological and economic impacts of management options by developing Forecasts and risk assessments. (Impacts)
5. Adaptive management and management evaluation. (Response)

(Return to 1.)

Drivers are considered large-scale anthropogenic and earth system phenomena that act through specific pressures to influence ecosystems. Examples of drivers include the increasing demand for seafood, increased human populations and their disproportionate migration to coasts, and long-term climate change affecting the atmosphere and oceans.

Pressures are the specific agents acting as a result of the drivers that affect ecosystems. For example, increase demand for seafood drives fishing effort, prices, and imports. Increasing human populations at the coasts generate higher levels of pollution and result in lower habitat quality, and global change may result in warmer temperatures and less sea ice.

States are various measures of current ecosystem conditions, such as the number of fishery stocks that are over fished, the average nutrient loads in coastal waters and the average water temperatures. Often these state variables are measured relative to some management imposed standards (e.g., through various laws).

Impacts are the consequences of the observed state of the system usually expressed in human terms such as total net benefits (or those foregone when ecosystems are degraded). They can also be expressed in other currency such as jobs, recreational opportunities, and other impacts humans care about. We envision IEAs to incorporate a risk assessment module to evaluate the risks and consequences of not meeting prescribed management targets as articulated in the selected set of state variables.

Last, the *response* part of DPSIR evaluates how the ecosystem state variables respond to the various management actions implemented. By iterating this model it is possible to build an empirical and modeling-based understanding of how the ecosystem responds to human pressures and to support adaptive learning and management schemes that achieve ecosystem objectives.

DESCRIBE ANY NOTABLE RESULTS, OUTCOMES OR IMPACTS TO DATE, IF ANY:

NOAA currently conducts several activities that will support future IEA development, and several small-scale integrated products are produced throughout the agency both routinely and on an ad-hoc basis. NOAA has not yet produced a full-scale regional IEA. This is a new product line envisioned for the agency.

PERFORMERS/OTHER PARTNERS (FEDERAL, STATES, OR LOCAL):

Federal: NOAA, EPA, NSF, and others
All Coastal States
Non-governmental Organizations:
Foundations such as Packard Foundation, Moore Foundation
COMPASS
University researchers
Others

PROJECT PERIOD:
NOAA plans to develop 8 regional IEAs over the next several years.

FUNDING LEVELS (CURRENT OR PROPOSED):
Approximately half of NOAA's $1.2 billion Ecosystems Programs will support data collection and integration efforts that will be used in IEA development. Additional funds are being sought.

TITLE OF PROJECT OR PROGRAM:
SERVIR: A Regional Visualization and Monitoring System for Improved Environmental Decision Making in Mesoamerica

AGENCY:
National Aeronautics and Space Administration (NASA); United States Agency for International Development (USAID)

PROJECT/PROGRAM DESCRIPTION:
SERVIR is a regional visualization and monitoring system for the nations of Central America that integrates satellite and in situ observations with environmental models for scientifically-based decision making by managers, researchers, students, and the general public. SERVIR addresses the nine societal benefit areas of the Global Earth Observation System of Systems (GEOSS): disasters, ecosystems, biodiversity, weather, water, climate, health, agriculture, and energy. For example, SERVIR can be used to monitor ecological changes and severe events such as forest fires, red tides, and tropical storms. In addition, SERVIR is developing forecasting tools for ecosystem change, as well as for weather and climate events.

SERVIR headquarters are located at the Water Center for the Humid Tropics of Latin America and the Caribbean (CATHALAC) in the Republic of Panama. A test bed and rapid prototyping SERVIR facility is managed by the NASA Marshall Space Flight Center at the National Space Science and Technology Center in Huntsville, Alabama.

DESCRIBE ANY NOTABLE RESULTS, OUTCOMES OR IMPACTS TO DATE, IF ANY:
The bilingual SERVIR website (http://servir.net/ or http://servir.nsstc.nasa.gov) provides free and open access to:

1. Satellite and Other Geospatial Datasets
    - Users search, browse, and download geospatial data and metadata
2. Interactive Online Maps
    - Users view live maps from dozens of Web Map Services
    - Users observe, animate, and download near real-time satellite feeds of regional weather and ecological conditions
3. Thematic Decision Support Tools
    - Users are accessing near real-time updates on fires, floods, red tides, and severe weather conditions (e.g., SERVIR was the central point for distributing international satellite imagery during and after Hurricanes Dean and Felix this hurricane season)

- Users browse from a selection of customized regional climate change, land cover, and ecological data products (e.g., leaf area index, land surface temperature, and fraction of absorbed photosynthetically active radiation)

4. 3D Interactive Visualizations
   - Users compare real-time visualizations of weather and other phenomena

In summary, utilizing SERVIR's flagship products (the SERVIR Data Portal, Real-time Image Viewer, and the SERVIR-VIZ visualization tool), users can search, browse, download, and visualize information from a variety of national, regional, and global geospatial sources addressing disasters, ecosystems, weather, climate, water, health, and other key thematic areas. The SERVIR team at CATHALAC is also equipped to prepare custom analyses, visualizations, GIS implementations, and educational products and services. An online User's Manual page gives more information about the website.

PERFORMERS/OTHER PARTNERS (FEDERAL, STATES, OR LOCAL):

SERVIR implementing agencies include NASA, CATHALAC, USAID, the Central American Commission for Environment and Development (CCAD), the World Bank, and the United Nations Environmental Programme (UNEP-ROLAC). Private sector Partners include: Cable and Wireless Panama, EGE Fortuna S.A., The Nature Conservancy, and the Institute for the Application of Geospatial Technology at Cayuga Community College, Inc. Other SERVIR key partners can be found on the SERVIR webpage under "Partners."

PROJECT PERIOD:

NASA is currently funding SERVIR for five years through 2008. USAID has been funding SERVIR at a comparable level over the same time period and has now taken a lead role in providing U.S. Government support to SERVIR.

FUNDING LEVELS (CURRENT OR PROPOSED):

NASA five-year funding for SERVIR totals approximately $3 million dollars while USAID funding has exceeded $3.5 million.

TITLE OF PROJECT OR PROGRAM:
Dynamics of Coupled Natural and Human Systems (CNH) http://www.nsf.gov/pubs/2007/nsf07598/nsf07598.pdf

AGENCY:
National Science Foundation (NSF)

PROJECT/PROGRAM DESCRIPTION:
The Dynamics of Coupled Natural and Human Systems (CNH) is a multi-directorate Program that promotes quantitative, interdisciplinary analyses of relevant human and natural system processes and complex interactions among human and natural systems at diverse scales. This competitive grant program is conducted jointly by three NSF directorates (Biological Sciences; Geosciences; and Social, Behavioral, and Economic Sciences), and beginning in 2008, in partnership with the Forest Service of the U.S. Department of Agriculture (USDA). Two additional NSF directorates (Engineering, and Education and Human Resources) and two offices (Office of International Science and Engineering, and Office of Polar Programs) participate on a less formal basis. CNH is a direct successor of the special competition on the part of the Biocomplexity in the Environment special competition on the Dynamics of Coupled Natural and Human Systems that was conducted from 2001 through 2005. CNH aims to support basic research and related activities that enhance fundamental understanding of the complex interactions within and among natural and human systems. The CNH competition promotes quantitative, interdisciplinary analyses of relevant human and natural system processes and complex interactions among human and natural systems at diverse spatial, temporal, and organizational scales. CNH seeks to advance basic knowledge about the system **dynamics**—the processes through which systems function and interact with other systems. Competitive proposals will focus on both **natural AND human systems** that are relevant to addressing the questions posed. Projects must also examine the full range of **coupled** interactions and feedbacks among relevant systems.

PERFORMERS/OTHER PARTNERS (FEDERAL, STATES, OR LOCAL):
Partnership with the Forest Service of the U. S. Department of Agriculture, starting fiscal year 2008.

PROJECT PERIOD:
2001 to present

FUNDING LEVELS (CURRENT OR PROPOSED):

Estimated Number of Awards: 7 to 12

Anticipated Funding Amount: Award sizes range from roughly $500,000 to no more than $1,500,000. The anticipated total in FY 2008 is $9,000,000. This total is for awards to be made annually, pending availability of funds.

Briefly describe the objectives of the program/project and how it deals with some of the core needs and scientific questions related to the use of the ecosystems concept in sustainable resource management or sustainable development.

Research funded by this program is expected to contribute to enhancement of theory within and across relevant fields. The team of researchers should include expertise from the natural sciences (biological sciences, geosciences, and/or physical sciences) and human sciences (social sciences, behavioral sciences, and/or engineering). Involvement of individuals with expertise in quantitative approaches and in education is also expected.

In addition to basic new knowledge and enhanced theory regarding the complex ways that people and natural systems interact, CNH seeks to develop the capabilities of people and tools needed to advance these areas of research in the future. CNH seeks to foster and develop new interdisciplinarity by bringing members of disparate disciplines into teams, and by developing new methods and expertise. In the process, the next generation of researchers will learn to work in diverse teams, cross disciplinary boundaries, and use advanced sensing and monitoring, communication and information technologies to work across many scales of time and space. A global perspective is encouraged in all proposals. Wherever appropriate and practical, specific international collaborations and networks for research and education are encouraged.

DESCRIBE ANY NOTABLE RESULTS TO DATE, IF ANY:

The funded projects described below provide a flavor of the projects that this Program fosters:

Award Abstract #0508028. BE/CNH: Understanding Linkages Among Human and Biogeochemical Processes in Agricultural Landscapes. P.I. Laurie Drinkwater, Cornell University

Humans have profoundly altered global cycling processes at multiple scales. Current estimates suggest human activities have doubled the amount of biologically active nitrogen on a global basis, with agriculture accounting for 75 percent of the human-derived nitrogen. A complex set of environmental and socio-economic factors influence agricultural fertilizer management practices. Linkages among socioeconomic and ecological subsystems are recognized as crucial in efforts to pursue sustainable ecosystem management and improve nitrogen-use amazon.com.

Billing Address:

Joseph Ferraro
132 E LINCOLN AVE
ROSELLE PARK, NJ 07204-1706
United States

Your order of September 14, 2010 (Order ID 102-7551191-5334632)

Qty.	Item
	IN THIS SHIPMENT
1	**Transitioning to Sustainability Through Research and Development on Ecosystem Services and Biofuels: Workshop Summary** Paperback **(** P-2-E3F2 **) 0309119820** **0309119820**

Subtotal
Shipping & Handling
Order Total
Paid via Visa
Balance due

This shipment completes your order.

Have feedback on how we packaged your order? Tell us at www.amazon.com/packaging.

143/D832zJGYR/-1 of 1-//1XSP/std-n-us/5671644/0915-09:00/0915-03:56/ecrump Pack Typ

pping Address:
ph Ferraro
E LINCOLN AVE
SELLE PARK, NJ 07204-1706
ted States

Item Price	Total
$31.75	$31.75

$31.75
$3.99
$35.74
$35.74
$0.00

http://www.amazon.com

For detailed information about this and other orders, please visit Your Account. You can also print invoices, change your e-mail address and payment settings, alter your communication preferences, and much more – 24 hours a day – at http://www.amazon.com/your-account.

Returns Are Easy!

Visit http://www.amazon.com/returns to return any item – including gifts – in unopened or original condition within 30 days for a full refund (other restrictions apply). Please have your order ID ready.

Thanks for shopping at Amazon.com, and please come again!

efficiency. Disconnections between human and natural subsystems must be addressed as well as disconnections within the component subsystems. Within the human realm, those who pollute do not pay the costs associated with resource degradation. Likewise, the biophysical system that has evolved as a result of high-input industrial agriculture is fraught with ecological disconnections. For example, uncoupling of carbon and nitrogen cycles is a defining trait of agricultural systems and is the root cause of the leakiness of these systems. On average, 45-55 percent of fertilizer nitrogen applied is lost to the environment. The goal of this research project is to understand how interactions among social and biophysical subsystems impact on carbon and nitrogen cycles in intensively managed agricultural landscapes at multiple scales. This project has implications for coupled human-natural systems theory and methodology, social systems theory, and environmental policy and will also have practical outcomes that are relevant to the development of agricultural and resource-management policy.

Award Abstract #0508002, BE/CNH: Urban Landscape Patterns: Complex Dynamics and Emergent Properties. P.I., Marina Alberti, University of Washington

Urban development in the United States is profoundly changing landscape patterns and biodiversity and is simultaneously affected by these changes. Little is known about the interactions between patterns and processes in human dominated landscapes, however. One of the least understood aspects of urban landscape dynamics is the way in which local interactions of humans and biophysical processes generate the landscape patterns of metropolitan regions. Studying the relationships between these interactions and the resulting urban landscape patterns is critical for planning and managing urban growth in ways that minimize the ecological impacts on ecosystems while sustaining economically and socially viable urban communities. This research project will examine urban landscapes as emergent phenomena that result from local interactions of human agents, real estate markets, built infrastructure, and biophysical factors such as land cover, geomorphology, and natural disturbance regimes to develop a theory of urban landscape dynamics. This study will employ complex-systems, patch-dynamics, hierarchical-theory, and agent-based modeling approaches to study coupled human-natural dynamics and empirically test this approach in two different bioregions (Seattle and Phoenix). The models will be developed and used to test hypotheses regarding emergent properties of urban landscapes and to enhance basic understanding of human-ecological interactions in urban landscapes across scales. Development of a better understanding of complex human-ecological dynamics leading to development patterns such as urban sprawl will contribute to the advance of biocomplexity science. The findings will also aid planning and management of urban regions by providing simulation tools to assess the ecological impacts and feedback of alternative strategies for urban development and ecological conservation.

Award #0505094, BE/CNH: Biodiversity Dynamics and Land-Use Changes in the Amazon: Multi-Scale Interactions Between Ecological Systems and Resource-Use Decisions by Indigenous Peoples. P.I. Jose Fragoso, University of Hawaii

Debate surrounding resource use and conservation by indigenous peoples has shifted away from tests of the "noble savage" hypothesis of the 1970s and 1980s towards analysis of the multiple social, economic, and biological factors that affect the sustainability of resource use. Hunting practices in particular among many indigenous groups are probably strongly regulated by internal controls, based on a combination of spiritual beliefs (cosmology), social rituals, and natural history knowledge. This research project will test the fundamental hypothesis that retention of traditional practices and cosmology by indigenous societies buffers them against the process of integration into the national society, thereby preventing biodiversity and ecosystem degradation by the indigenous societies themselves. Socioeconomic data, wildlife data, and remotely sensed data will be collected, integrated, and analyzed within a geographic information system. In addition to a better understanding of human-biodiversity linkages in indigenous areas, outcomes of this project will include (1) educational materials for the Macuxi and the institutions that work with them, (2) a distance-linked graduate seminar in which students collaborate across departments, campuses and disciplines, (3) broadening of the participation of women and minority students in science, and (4) enhancement of the infrastructure for science by linking institutions with different areas of specialty into a teaching and research network that will benefit students who would normally have access only to their own institution. This project will contribute to the development of effective development policies and biodiversity conservation and will help provide theoretical background for coupled human-natural systems in the subsistence or semi-subsistence societies that characterize much of tropics. The results will be particularly germane for the ongoing debate on the role of "people in parks" and on the contribution that indigenous peoples will make to biodiversity conservation worldwide. The geographical location of this study is significant unto itself. Roraima covers a large portion of the unstudied and largely unmanaged high diversity Guiana Shield forest-savanna transition. For this key ecological area, the future of biodiversity lies in the hands of indigenous peoples. This study will provide insights into the internal cultural dynamics of indigenous societies and how they influence, and are influenced by, biodiversity patterns and ecosystem function. The results will have important implications for human-environment interactions in Raposa and elsewhere where indigenous peoples retain an important presence.

D

Descriptions of Agency Activities Presented at the Forum on Biofuels and Sustainability[1]

[1]Presentations are available online at http://sustainability.nationalacademies.org/Forum.shtml

TITLE OF PROJECT OR PROGRAM:
Renewable Energy Assessment Program (REAP)

AGENCY:
USDA Agricultural Research Service

PROJECT/PROGRAM DESCRIPTION
Domestic ethanol production is a strategy for reducing dependence on imported energy and release of greenhouse gases from use of fossil-energy-derived motor vehicle fuel. Federal and state governments are encouraging the use of ethanol. Initially energy crops, such as switchgrass, willow, and poplar, were targeted as sources of bio-energy, recently crop residues, especially corn stover and wheat straw, have been identified as a source of cellulosic biomass. However, the amount of crop residue needed to protect soil from erosion and to sustain soil organic carbon (SOC) stores constrains residue removal for bio-energy. Research over the past century has shown conclusively that crop production practices result in loss of SOC. Typically loss of SOC has detrimental effects on soil productivity and quality. Our objectives are to determine the amount of residue needed to protect the soil resource, compare economic implication based on the value of stover as bio-energy and C source, and provide initial harvest rate recommendations and guidelines. Products from this work will be 1) guidelines for management practices supporting sustainable harvest of residue, 2) algorithm(s) estimating the amount of crop residue that can be sustainably harvested, and 3) decision support tools and guidelines describing the economic trade-off between residue harvest and retention to sequester soil C. Delivery of this knowledge and these products to farmers and the biomass ethanol industry will promote harvest of stover and crop residues in a manner that preserves the capacity our soil to produce food, feed, fiber, and fuel.

DESCRIBE ANY NOTABLE RESULTS, OUTCOMES OR IMPACTS TO DATE, IF ANY:

PERFORMERS/OTHER PARTNERS (FEDERAL, STATES, OR LOCAL):
ARS laboratories in Lincoln, NE; Mandan, ND; Auburn, AL; St. Paul, MN; Ames, IA; Ft. Collins, CO; Pendleton, OR; Morris MN; W. Lafayette, IN.

PROJECT PERIOD:
Start Date: Jun 01, 2006 End Date: May 31, 2011

FUNDING LEVELS (CURRENT OR PROPOSED):

Note this is not a separately funded project. Participating scientists contribute their time and resources out of their base funding to projects falling under ARS Soils, Bioenergy and Global Change National Programs.

TITLE OF PROJECT OR PROGRAM:
Future Midwest Landscapes Study

AGENCY:
EPA, Office of Research and Development

PROJECT DESCRIPTION

Geographic focus

The Future Midwest Landscapes (FML) Study is a place-based component of the EPA's Ecological Research Program (ERP), and shares the ERP's goals of conserving ecosystem services by providing support and tools that enable decision-makers, from national to local scales, to recognize and account for these services (see ERP summary). The FML will focus on the valuable and productive agroecosystems of the Midwest, using an alternative futures approach to model varying trajectories of landscape change and to evaluate implications of these changes for ecosystem services and, thus, human well-being. Because the rapid growth of the biofuels industry currently is the dominant driver of landscape change in the region, the future scenarios to be examined will center on biofuels development, and the study area (see red outline in Figure 1) will include locations of current and projected change.

The FML *project goals* are as follows:

- **Understand** how current and projected land uses affect the **ecosystem services** provided by Midwestern landscapes
- Provide **spatially explicit information** that will enable EPA Regions and Programs to articulate sustainable approaches to environmental management in the Midwest
- Develop **web-based tools** depicting alternative futures so users can evaluate trade-offs affecting ecosystem services.

The alternative-futures *research approach* will involve the following steps (Liu et al.):

1. Scenario Definition

Stakeholder meetings will explore values related to alternative futures for the Midwest. Two distinct types of future scenarios will be created, differing in how landscape change is approached:

Forecasting ('what-if') scenarios will project the landscapes that would be expected to result from a divergent set of potential energy and agricultural policies. These landscapes will be analyzed to evaluate the regional-scale impacts on ecosystem services as well as implications to national-scale issues such

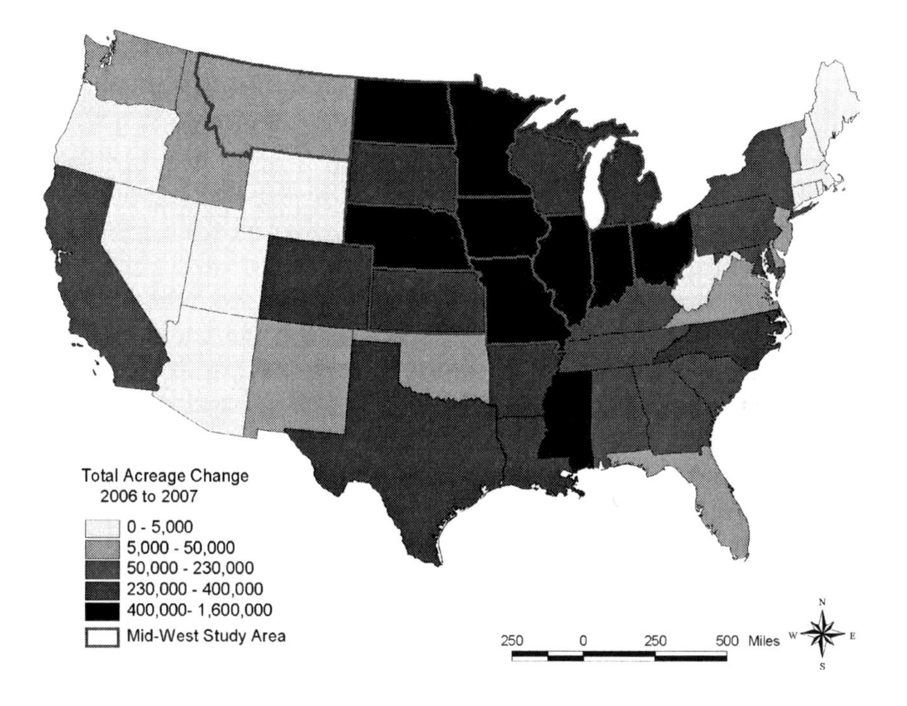

as chemical runoff into the Mississippi River and Lake Erie and loss of critical migratory bird habitat.

Backcasting ('how-could') scenarios will entail a suite of landscapes that seek to emphasize all ecosystem services by placing crops according to soil erodability and productivity, opportunities to provide wildlife habitat, protection of drinking water, etc. These scenarios will help users to explore what is possible and to identify goals at the local or regional level.

2. Scenario Construction

Future economic drivers will be modeled for each forecasting scenario using the FAPRI system (agricultural sector) and MARKAL (energy sector). Maps of projected landscape change corresponding to each (forecasting and backcasting) scenario will be produced.

3. Scenario Analysis

Using models of agronomy, hydrology, biogeochemistry and habitat suitability, the following ecosystem services (and well-being effects) will be estimated, to the extent feasible, and compared to baseline conditions:

- Soil productivity (affects food and energy security)
- Carbon balance (affects climate)

- Predator refugia (controls pests)
- Hydrologic and water quality regulation (affect water supply, flooding, downstream aquatic ecosystems, recreation)
- Wildlife habitat and other natural areas (affect biodiversity and recreation)
- Air quality (affects health)

4. Scenario/Risk Assessment

Web-based tools will be used to visualize and present results. The landscape analysis methods developed for the FML Study will be implemented as a web-based environmental decision toolkit (EDT), similar to other toolkits previously created under EPA's Regional Vulnerability Assessment Program (ReVA). We anticipate that the future FML-EDT will allow users to compare alternative Midwestern futures by examining trade-offs—that is, changes in the provision of a wide variety of ecosystem services—at both local and regional scales.

For local-scale decision-makers, we will also investigate the feasibility of incorporating ecosystem services into existing software applications that support decisions at watershed or farm scales.

5. Risk Management

By initially engaging potential users in EPA's regional and program offices, USDA, and farm, industry and conservation associations, we expect to be able to maintain stakeholder involvement and directly support uses of these tools.

DESCRIBE ANY NOTABLE RESULTS, OUTCOMES OR IMPACTS TO DATE, IF ANY:

To date we have held planning meetings in EPA Regions 5 (Chicago) and 7 (Kansas City). We are in the process of establishing cooperative agreements

PERFORMERS/OTHER PARTNERS (FEDERAL, STATES, OR LOCAL):

- EPA Region 7 (Kansas City)
- EPA Region 5 (Chicago)
- Midwest Spatial Decision Support System Partnership
- Kansas State University
- Iowa State University

PROJECT PERIOD:

Start Date: Jan. 2005 End Date: Dec. 31, 2014

FUNDING LEVELS (CURRENT OR PROPOSED):

20 scientist FTE and $1M extramural support annually (proposed)

TITLE OF PROJECT OR PROGRAM:
Bioenergy from Forests- Moving Science to Practice and Issues of Sustainability

AGENCY:
U.S. Forest Service

PROJECT/PROGRAM DESCRIPTION:
The forestry sector in the United States has been a major source of renewable energy and currently produces over half of the renewable energy generated in the US. This industrial sector is the logical platform for expanding forest based energy in the form of heat, power, and transportation fuels. As opportunities expand potential for wood-based energy, there are sustainability issues that fundamentally touch on meeting the needs of people today without compromising our ability to meet future generation's needs, as well as careful consideration of the environmental, social, and economic dimensions of forest-based biofuels. In order to address the topic of sustainability and bioenergy from forests, it is important to consider the full continuum beginning with forests growing on the landscape; decisions about when, where, how and why we remove biomass to produce energy and co-products; and advances in efficiency and effectiveness of converting biomass to energy in what will likely be a carbon constrained future.

As each of these segments of the continuum is discussed, examples are presented of activities currently underway in both R&D and in practice that demonstrate how the forestry sector is addressing the issues that relate to sustainability. Because sustainability also needs to be discussed in terms of scale, examples are presented that show how forest bioenergy contributes to community, regional, national, and international sustainability.

Growing healthy and productive forests on the landscape has been a focus of R&D for nearly 100 years. Studies that evaluate long term site productivity have provided insight into repeated removal of biomass and effects on nutrient cycling, and studies that evaluate the needs of wildlife and the hydrological cycle have provided insight into how biomass removal strategies either impact or can improve these other values. Understanding trends and conditions of the forest resource in order to provide accurate feedstock assessment information for local, regional, and national planning is critical because wood used for energy is available for biofuels only in the context of other forest products and values. For example, plantation forests in the Southern US were planted primarily for the pulpwood market, but as this market shifts more to using pulp imports, then management of these forests may shift to respond to a biofuels market when conducting forest thinning. Reducing fuel loading on forests in order to reduce fire risk around communities in the west also provides a local source of biomass for smaller scale bioenergy applications.

Research over the past 25 years has provided a valuable foundation for planting fast growing forest biomass (such as hybrid poplar) to meet demands of a growing bioenergy market, and as part of a diversified biomass crop feedstock strategy. Continuing the period assessments that forms the basis for the National Report on Sustainable Forestry provides the ties and linkages among the various dimensions of managing forests across landscapes, how forests are being impacted by climate change, and the need for adaptation to and mitigation of these effects.

Decisions about when, where, how, and why we remove biomass to produce energy and co-products occur in the context of land owner/land management objectives. The objectives of public land management are described in their land and resource management plans and often include restoring ecological processes and functions, improving resilience to climate stressors, improving habitat, reducing risk of wildfire, mitigating impacts of fire, insects and disease, and producing forest products for society. Private forest land owners have economic and social objectives along with interests in maintaining healthy and productive forests for a variety of purposes. Economic incentives to maintain "working" forests on the landscape rather than convert the private land to other uses is a critical part of the trade-off considerations. Vast areas of forests were historically cleared in order to develop productive agriculture and grow our cities. The potential for reestablishing forest energy crops on marginal crop land or as longer term conservation areas will likely be part of the future dialog to meet land owners needs and provide longer term options for the country. Biodiversity considerations will help guide where expanded wood energy crops can provide multiple benefits, or may limit how and where they are planted.

The ability to implement projects with a significant component of biomass is challenging primarily due to the high cost of harvesting and transportation of large volumes of lower value materials. R&D has provided improved harvesting and transportation strategies such as biomass bundlers, and roll off bins to help address transportation costs. This is the highest cost center and is currently only economical when higher value products help pay for handling the lower value biomass. Economic studies have demonstrated a radius of about 50 miles for economically transporting biomass to a bioenergy facility, and these historically have been for power production. Most biomass that is used for bioenergy today are residues that are already at a mill and are integrated to provide heat and power, adding additional value by residue disposal.

Urban areas also provide woody biomass residues that can be used for bioenergy. The city of St. Paul, MN, heats and cools, and provides electricity to the downtown area from a central facility using clean wood materials from urban tree trimming, storm damage and forest land clearing. Matching the scale of the bioenergy project to the needs of the community and to the forest biomass supply is a critical part of making sure the bioenergy is sustainable. Understanding the full carbon cycle of using hybrid poplar trees as a feedstock for this facility has been one of the dimensions of R&D and management collaborative efforts.

A growing area of potential trade-off in forest management activities is to offset emissions generated from burning of slash piles and instead burn the material in controlled combustion to produce bioenergy. Currently this is feasible for co-firing with coal or other fossil fuels or as dedicated biomass boilers. As the concept of a biorefineries is demonstrated at commercial scales R&D continues to evaluate this potential trade-off, including exploring how reducing hazardous fuels may result in smaller and less intense wildfires, and may result in reduced emissions from wildfire. This is an active area of R&D discussed in the separate Alder Springs project in California. A 10-year stewardship contract in Arizona demonstrates how collaboration can help bridge interests and provide a landscape level strategy that is socially, economically, and ecologically acceptable to a wide group of stakeholders.

The final major focus of R&D is **advance efficiency and effectiveness of converting biomass to energy in a carbon constrained future.** When woody materials are converted to pulp to make paper and other wood fiber products there are by-products and co-products that can be used for biofuels. Currently the lignin is burned to generate heat and power for the pulp mill. However, both the lignin, and the 5- and 6-carbon sugars in the wood pulp are the essential building blocks for biofuels. R&D is providing economic analysis to determine if forest biorefinery concepts are feasible, evaluate alternative production models, and is developing pricing and financial models. R&D is improving pentose fermentation, pretreatment of cellulose, extent to which co-products can be produces prior to pulping, and developing products from lignin. Industrial application of the new methods involves scaling up and testing with industrial partners.

Pyrolysis of woody material into a high energy-density-intermediate product is the focus of R&D efforts that could also help with more economical shipping. New gasification technology to increase intermediate gas production and reduce char formation provides opportunities for using a more diverse feedstock mix.

For each conversion technology being studied, there is an associated feedstock characterization component in order to match the forest biomass material with compatible processes for bioenergy. Sustainability requires careful consideration of the local nature of woody biomass, and how it fits into a large scheme of cellulosic crops and residues that may be available for biofuels.

Finally, R&D using life cycle analysis is demonstrating the net energy impacts considering the full continuum beginning with growing forests through energy production in order to fully consider benefits and trade-offs of various feedstocks for bioenergy, and for evaluating biofuels as compared to the fossil fuels they replace. There has been extensive research that demonstrates the life cycle benefits of using wood products when compared to steel and concrete for many types of construction. Bioenergy currently used in processing of wood products is one of the reasons. Similar analysis of the full carbon cycle for bio-energy from forests is underway.

PERFORMERS/OTHER PARTNERS (FEDERAL, STATES, OR LOCAL):

Forest Service R&D works with a wide variety of partners to accomplish this work. Partners include other federal agencies such as the Department of Energy, Environmental Protection Agency, researchers in other USDA Agencies, and a wide variety of university scientists. Forest genetics and conversion technology have also included private industrial partners. Implementing research and demonstration projects on the ground frequently involves National Forest System and other public land managers, or State Forestry and Environmental Agencies. Local collaborative groups in many areas are helping build consensus about how bio-energy projects fit into the local and regional business options for achieving land management objectives. Governor's associations such as the Western Governors are building regional strategics and options for incorporating bioenergy into a "clean and diversified energy" vision.

TITLE OF PROJECT OR PROGRAM:
Integrated Mapping of Biofuels Feedstock Production and Transportation for Systems Visualization and Optimization

AGENCY:
U.S. Department of Transportation

PROJECT/PROGRAM DESCRIPTION:
One unintended consequence of a large scale shift from fossil fuels to biofuels may be localized undercapacity of the Nation's freight transportation system. Because feedstocks are roughly an order of magnitude less energy dense than fossil fuels, transportation of the required large volumes of these feedstocks over long distances may be unrealistic from both a transportation capacity and a lifecycle energy perspective. In addition, because ethanol is hydrophilic, it is not possible to use current pipelines for transport. Instead, it is necessary to rely on barge, train, and trucks. Depending on where feedstocks are produced and processed and where the resulting biofuels are used, some roads, barge routes, and rail lines may see a significant increase in freight ton-miles.

A synthesis of results from Sun Grant and DOE research is being used to populate a suite of GIS tools for future biomass development planning to address these questions in the context of other lifecycle costs and impacts for biofuels. The Sun Grant Centers are currently in the process of integrating multi-modal transportation maps with agricultural feedstock production maps. Among other applications, the integrated GIS tools will be used to evaluate the potential impacts of the transportation infrastructure on biomass energy and materials development and visa versa. These tools will enable focused trade-offs for different feedstock production options and geographic distribution of processing and pre-processing options. They will enable a rigorous trade-off between centralized and distributed production and processing options as the feedstocks and production technologies mature.

DESCRIBE ANY NOTABLE RESULTS, OUTCOMES OR IMPACTS TO DATE, IF ANY:
N/A

PERFORMERS/OTHER PARTNERS (FEDERAL, STATES, OR LOCAL):
Federal funding and research partners: US Department of Transportation, US Department of Energy, US Department of Agriculture.

Charter principal institutions: Sun Grant Association, Cornell University, Oklahoma State University, Oregon State University, South Dakota State University, and University of Tennessee.

Participants are from the above public and private institutions.

PROJECT PERIOD:

The project was started in 2006.

FUNDING LEVELS (CURRENT OR PROPOSED):

$0 currently.

TITLE OF PROJECT OR PROGRAM:
Biobased R&D

AGENCY:
U.S. Department of Transportation, Research and Innovative Technology Administration

PROJECT/PROGRAM DESCRIPTION:
In 2006, the Sun Grant Initiative (http://www.sungrant.org) and the National Biodiesel Board (NBB) (http://www.biodiesel.org) received funding from the US DOT to conduct biobased R&D with a focus on addressing DOT priorities. This project is coordinated by the 5 regional sun grant universities and the NBB and is addressing a wide range of research items necessary to develop biofuels as a significant, sustainable contributor to the national energy needs. The projects will address several high-level topic areas including:

- Sustainable feedstock development
- Infrastructure analysis
- Economic analysis
- Efficient conversion technologies, include those emplacing local and distributed production
- Emissions and fuel quality testing

This project is of great relevance due to increasing interest in reducing transportation contribution to greenhouse gas emissions, and the need to effectively plan and invest in energy transport infrastructure.

The funding recipients are working closely with industry and agriculture partners to facilitate commercialization or large scale demonstration of successful laboratory and university based research. The USDOT is working with the grantees to maximize the connection to DOT research priorities and help ensure that many of the systems analysis requirements are conducted. These include infrastructure requirements for feedstock and fuel transport, life-cycle and environmental impact analysis. Although applied laboratory research is important and a component of the program emphasis is placed on technologies that can be successfully integrated into the transportation/energy system.

DESCRIBE ANY NOTABLE RESULTS, OUTCOMES OR IMPACTS TO DATE, IF ANY:
Selection of individual projects and partners.

PROJECT PERIOD:
November 2006 through December 2014

PARTNER FUNDING LEVELS (CURRENT OR PROPOSED):
$50,000,000 authorized
$41.66 to the 5 sun grant regional universities
$8.33 Million to the National Biodiesel board

AGENCY:
Department of Energy (Office of the Biomass Program)

PROJECT/PROGRAM DESCRIPTION:
Biomass research has been a cornerstone of DOE's renewable energy research, development and deployment efforts over the last 25 years. In order to encourage the economic livelihood of a thriving biofuel industry, the Office of the Biomass Program (OBP) at the Department of Energy supports research and development aimed at assessing the impacts of biofuels on the environment, including impacts to land, water, and air from energy production and use. Included in this mission is a goal to substantially reduce greenhouse gas emissions by accelerating the adoption of renewable energy technologies.

A clear driver of the OBP's activities is the President's goal to increase the use of biofuels and other alternative fuels in the transportation sector to replace 20% of the gasoline demand in the United States by 2017 (referred to as the "20 in 10" goal). Meeting this goal will require: significant and rapid advancements in biomass feedstock and conversion technologies; availability of large volumes of sustainable biomass feedstock; demonstration and deployment of large-scale integrated biofuels production facilities; and biofuels infrastructure development efforts. In addition, the existing agricultural, forestry and commercial sectors will be making the decisions to invest in biomass systems—from shifting land use, to building capital-intensive biorefineries, to establishing the infrastructure and public vehicle fleet for ethanol distribution and end use—in the context of economic viability (including as it relates to environmental sustainability) and the needs of the marketplace.

Research and Development Priorities related to Biofuels and Sustainability. OBP's R&D has led the effort to develop technology necessary to sustainably produce, harvest, and convert a variety of biomass feedstocks, as well as to deploy the resulting biofuels. Core R&D on feedstock production and logistics and biomass conversion technologies is conducted to develop the scientific and technical foundation that will enable the new bioindustry. OBP is looking to advance science in these areas through important collaborations with the DOE Office of Science Bioenergy Centers, the U.S. Department of Agriculture, land grant universities, and private industry. OBP has developed Regional Feedstock Partnerships to begin to realize the sustainability of the resource potential outlined in the Billion Ton Study. This approach facilitates the collaboration of industry, the agricultural community, state and local governments, and USDA and is expected to accelerate the resource readiness, as the cellulosic fuels industry emerges.

The core R&D of OBP is organized around the integrated biorefinery concept. The biorefinery helps deliver sustainable and environmentally sound contributions to power, fuels, and products demand while supporting rural economies. Key barriers relevant to this area include ensuring resource sustainability at levels large enough to support large-scale production facilities and maximizing the efficiency of conversion facilities to minimize costs. Energy production from biomass on a large scale will require careful evaluation of U.S. agricultural resources and logistics, as these will likely require a change in paradigm that will take time to implement. Current harvesting, storage and transportation systems are currently inadequate for processing and distribution of biomass on the scale needed to support dramatically larger volumes of biofuels production. Evaluating the current feedstock resource on a national level as well as the potential for future feedstock production in light of environmental constraints is part of OBP's focus.

Major Research and Development Program Areas

The primary program areas related to ecosystems services and sustainability within DOE OBP are (1) sustainable feedstock production, (2) sustainable harvest, and (3) sustainable biofuels production.

Sustainable feedstock production

Existing data on the environmental effects of feedstock production and residue collection are not adequate to support lifecycle analysis of biorefinery systems. The lack of information and decision support tools to predict effects of residue removal as a function of soil type, and the lack of a selective harvest technology that can evenly remove only desired portions of the residue make it difficult to assure that residue biomass will be collected in a sustainable manner. Until the residue issue is addressed, particularly with regard to corn stover, deployment of the Agricultural Residue pathway will be severely constrained. The production and use of perennial energy crops also raise a number of sustainability questions (such as water and fertilizer inputs, establishment and harvesting impacts on soil, etc.) that have not been comprehensively addressed.

A central focus of feedstock production efforts is to establish and maintain Regional Biomass Energy Feedstock Partnerships in collaboration with USDA, the Sun Grant Initiative universities, and other regional partners. Collaborating in this manner will be crucial to overcoming specific geographic issues of varying climatic conditions, soil types, water quality, and land usage. These regional partnerships are necessary for assessing and quantifying the feedstock resource base because of the diversity of regional feedstocks in terms of growth requirements, climatic differences, and infrastructure needs. Work will focus on:

1) Genetic improvement of crops including yield per acre as well as traits desirable for conversion to biofuels, an effort conducted primarily by USDA, DOE's Office of Science, land grant universities, and private companies;
2) Regional resource assessments of the types of biomass feedstocks that can be sustainably grown in specific locations across the U.S., including the development of a GIS-based bioenergy atlas;
3) Development of replicated field trials across regions to determine the impact of agricultural residue removal and to evaluate the feasibility of various energy crops; and
4) Analysis of the sustainability of producing specific biomass feedstocks, an effort being addressed by the Regional Feedstock Partnerships, USDA, and OBP analysis efforts.

The GIS tool will serve as a spatially referenced database of current and potential feedstock availability and associated environmental and industrial variables to be used in the analysis of future economic and environmental sustainability related to feedstocks. This tool will also serve as a decision support system to inform the location of new feedstock production and processing facilities and to evaluate the resulting contribution potential of biofuels to the "20 in 10" goal (and beyond) that will sustain air and water resources of quality and availability for desired uses.

Sustainable harvest

Current crop harvesting machinery is unable to selectively harvest desired components of biomass and address the soil carbon and erosion sustainability constraints. Biomass variability places high demand and functional requirements on biomass harvesting equipment. Current systems cannot meet the capacity, efficiency, or delivered price requirements of large cellulosic biorefineries, nor can they effectively deal with the large biomass yields per acre of potential new biomass feedstock crops. In addition, feedstock specifications and standards against which to engineer harvest equipment, technologies, and methods, do not currently exist.

A key to success is the ability to convert a wider variety of regionally-available biomass feedstocks and agricultural waste. The Department of Energy is working to establish Regional Biomass Energy Feedstock Partnerships that will identify local opportunities for feedstock development and ethanol production.

Sustainable biofuels production

While perhaps the greatest sustainability challenges to biofuel production lie within the feedstock production sector of the biomass-to-bioenergy supply chain, existing projects within OBP extend beyond feedstock production and harvest. Currently, a life cycle assessment (LCA) of the Advanced Energy Initiative is being performed for the 60 billion gallon 30x30 scenario (a scenario for supplying 30% of 2004 motor gasoline demands by 2030). The analysis covers the entire biofuels supply chain from feedstocks to vehicles. The four main areas addressed in the LCA are: land use and soil sustainability, water use impacts, air quality impacts, and greenhouse gas (GHG) emissions impacts. Also, the GREET model (Greenhouse gases, Regulated Emissions, and Energy use in Transportation) is being utilized for an analysis of water demand for biofuel production, energy and GHG emission benefit of biofuels. Included in this project is an expansion of the existing model to include corn ethanol, sugarcane ethanol, and flex-fuel vehicle (FFV) test results.

Estimated annual funding levels for research and development activities related to ecosystems services and sustainability:

FY 2007 and 2008 funding for biofuels research directly related to sustainability is approximately $5 million. Total Biomass Program Funding supporting all R&D efforts is approximately $150M in FY07.

E

Biographical Information Roundtable on Science and Technology for Sustainability

PAMELA A. MATSON (NAS) is Naramore Dean of the School of Earth Sciences and Goldman Professor of Environmental Science at Stanford University. Her current research interests include biogeochemical processes in forest and agricultural systems. Dr. Matson was the first to show that geographic variation in biogeochemistry of terrestrial ecosystems controls variation in the production of the important greenhouse gas N_2O. That discovery provided the foundation for her development of global budgets of natural and anthropogenic sources of this and other radiatively significant trace gases. Dr. Matson has served on numerous National Academies' committees, including the Board on Sustainable Development, the Committee on Research and Peer Review in EPA, the Board on Global Change, and others. She is President of the Ecological Society of America, a member of the Aspen Global Change Institute Advisory Board, and a member of the Institute of Ecosystem Studies Advisory Board. Selected publications include Ecosystem Approach for the Development of a Global Nitrous Oxide Budget; Agricultural Intensification and Ecosystem Properties; and Integration of Environmental, Agronomic, and Economic Aspects of Fertilizer Management. Dr. Matson received her B.S. in Biology from the University of Wisconsin – Eau Claire; her M.S. in Environmental Science from Indiana University; and her Ph.D. in Forest Ecology from Oregon State University.

EMMY SIMMONS recently retired from the position of Assistant Administrator for Economic Growth, Agriculture, and Trade at the United States Agency for International Development (USAID). Ms. Simmons has more than 30 years experience in international agriculture and economic development. Since 1997 she has served as USAID deputy assistant administrator in the former Bureau for Global Programs, Research and Field Support, where she headed the Center for Economic Growth and Agricultural Development. From 1994 to 1997, Simmons was senior program officer for USAID's mission in Moscow where she oversaw an aid portfolio of more than $1 billion. From 1991 to 1994, she served in USAID's regional office for east and southern Africa as supervisory program economist. Simmons also has served as supervisory agricultural officer for Mali and as regional agricultural advisor for West Africa, in addition to holding a number of supervisory positions in the Africa Bureau in USAID's Washington headquarters. She received her B.S. from the University of Wisconsin-Milwaukee; her M.S. in Agricultural Economics from Cornell University.

MATHEW ARNOLD is a co-founder of Sustainable Finance Ltd. He assists financial institutions and investors in understanding and managing environmental and social risks, and helps them identify environmentally superior investment opportunities. His current clients include Citigroup, JP Morgan Chase, LaSalle Bank, and the Global Environment Fund, a private equity investment company dedicated to environmental investments. He was Chief Operating Officer at the World Resources Institute, a sustainable development think tank. As COO, he was responsible for 140 employees, a $20-million budget and a $45-million endowment. He has served as an advisor on environmental and sustainability strategy for several other multinational corporations including DuPont, where he helped develop the organization's sustainable growth strategy, and BP, where he helped business unit leaders to align with the company's post-Amoco merger focus on greening the brand. In 1990, he founded the Management Institute for Environment and Business (MEB) to help business schools and corporations integrate environmental issues into business strategy. In 1996, MEB merged with the World Resources Institute. Prior to 1990, he held positions in marketing with IBM, in investment banking with Merrill Lynch Capital Markets, and in business development with Santa Fe Trading, Hong Kong. He is a member of the Board of Directors of Forest Trends, a market maker for ecosystem services. He holds an AB degree in Psychobiology from Harvard College, an M.A. in International Relations from the Johns Hopkins University, and an M.B.A. from the Harvard Business School.

ARDEN L. BEMENT, JR. (NAE), Director, National Science Foundation, joined NSF from the National Institute of Standards and Technology, where he had been director since 2001. As head of NIST, he oversaw an agency with an annual budget of about $773 million and an onsite research and administrative staff of about 3,000. Bement previously served as the David A. Ross Distinguished Professor of Nuclear Engineering and head of the School of Nuclear Engineering at Purdue University. He has held appointments at Purdue University in the schools of Nuclear Engineering, Materials Engineering, and Electrical and Computer Engineering, as well as a courtesy appointment in the Krannert School of Management. He was director of the Midwest Superconductivity Consortium and the Consortium for the Intelligent Management of the Electrical Power Grid. Along with his NIST advisory roles, Bement served as a member of the National Science Board from 1989 to 1995. He also chaired the Commission for Engineering and Technical Studies and the National Materials Advisory Board of the National Research Council. Additionally, he was a member of the Space Station Utilization Advisory Subcommittee and the Commercialization and Technology Advisory Committee for NASA and consulted for the Department of Energy's Argonne National Laboratory and the Idaho National Engineering and Environmental Laboratory. Prior positions include: vice president of technical resources and of science and technology for TRW Inc. (1980-1992); deputy under secretary of defense for research and engineering (1979-1980); director, Office of Materials Science, DARPA (1976-1979); professor of nuclear materials, MIT (1970-1976); manager, Fuels and Materials Department and the Metallurgy Research Department, Battelle Northwest Laboratories (1965-1970); and senior research associate, General Electric Co. (1954-1965). He has been a director of Keithley Instruments Inc. and the Lord Corp. and was a member of the Science and Technology Advisory Committee for the Howmet Corp. (a division of ALCOA). Bement holds an engineer of metallurgy degree from the Colorado School of Mines, a master's degree in metallurgical engineering from the University of Idaho, a doctorate degree in metallurgical engineering from the University of Michigan, an honorary doctorate degree in engineering from Cleveland State University, and an honorary doctorate degree in science from Case Western Reserve University. He is a member of the National Academy of Engineering.

MICHAEL D. BERTOLUCCI is the President of Interface Research Corporation (IRC), Chairman of the Envirosense® Consortium, Inc.—a not-for-profit organization concerned with Indoor Air Quality—and Senior Vice President of Interface, Inc., a billion dollar enterprise with over 5000 employees. With manufacturing sites in seven countries, Interface is the world leader in the sale of modular carpet tiles and commercial interior fabrics. The research arm, along with the various divisions' R&D laboratories provides fundamental technology to the overall enterprise. As president, Dr. Bertolucci leads not only this effort but also the parent's mission to become the first name in industrial ecology and to provide new technical solutions for Interface as it strives to reduce its footprint on the environment and to become sustainable. He serves on the boards of several not-for-profit organizations, such as the CEO Coalition to Advance Sustainable Technology (CAST), and the oversight committee of the National Research Council's Division on Earth and Life Studies (DELS). Prior to coming to the IRC, Mike spent six years as Vice President of Technology for Highland Industries, an industrial fabrics company, fifteen years in numerous research and development management posts with the General Electric Plastics Business Group, and four years in chemical research at Union Carbide Chemicals and Plastics. Dr. Bertolucci received his Ph.D. in Physical Chemistry from the California Institute of Technology, and his B.S. degree in Chemistry from San Jose State.

JOHN CARBERRY is Director of Environmental Technology for the DuPont Company in Wilmington, Delaware where he has been employed since 1965. He is responsible for recommendations on technical programs for DuPont based on an analysis of environmental issues. Since 1989, he has led this technology function in a transition to increasingly emphasize waste prevention and product stewardship while maintaining excellence in treatment. Externally, Mr. Carberry is a past Chair of the standing National Academy Committee on the Destruction of the Non-Stockpile Chemical Weapons, a founding member of the Green Power Market Development Group and of the Vision2020 Steering Committee, and a member of the NAE Committees on; Technologies for Sequestering CO_2, and Metrics for Documenting Progress in Global Change Research. Since 1990, John has served on four other National Academy Committees and has presented 30 lectures on environmental issues at 18 universities, given invited presentations at 63 public conferences worldwide and provided 21 literature interviews, or contributions. He holds a B.ChE. and an M.E. in Chemical Engineering from Cornell University and an MBA from the University of Delaware.

LESLIE CAROTHERS is President of the Environmental Law Institute. ELI is an independent, non-partisan education and research organization working to protect the environment by improving law, policy, and management. She has been a professional environmentalist for over 30 years. Before her election as ELI president in June 2003, she served for 11 years as Vice President, Environment, Health and Safety at United Technologies Corporation (UTC) in Hartford, a diversified manufacturer of products for the aerospace and building systems markets. She also served as Commissioner of the Connecticut Department of Environmental Protection from 1987-1991 and Senior Environmental Counsel for PPG Industries, a manufacturing company in Pittsburgh, from 1982-1987. She began her environmental career with the federal Environmental Protection Agency in the air pollution program in Washington in 1971 and later served as Enforcement Director, Deputy Regional Administrator, and Acting Regional Administrator of EPA's New England Region in Boston. In 1991, she was an adjunct lecturer on environmental regulation at the Yale School of Forestry and Environmental Studies. She is a past member and Chair of the Board of Directors of the Connecticut Audubon Society and the Environmental Law Institute and a past member of the Board of the Nature Conservancy (Connecticut Chapter). She currently serves on the Board of Directors of Strategies for the Global Environment (Pew Center on Global Climate Change). She is a graduate of Smith College and Harvard Law School and also holds a Masters Degree in environmental law from George Washington University.

WILLIAM CLARK (NAS) is the Harvey Brooks Professor of International Science, Public Policy and Human Development at Harvard University's John F. Kennedy School of Government. Trained as an ecologist, his research focuses on the interactions of environment, development and security concerns in international affairs. Clark serves on the scientific advisory committees for the Science and Technology for Sustainability Initiative, the International Human Dimensions Programme on Global Environmental Change and the Potsdam Institute for Climate Impacts Research. He is co-author of Adaptive environmental assessment and management (Wiley, 1978) and Redesigning rural development (Hopkins, 1982); editor of the Carbon dioxide review (Oxford, 1982); and coeditor of Sustainable development of the biosphere (Cambridge, 1986), The earth transformed by human action (Cambridge, 1990), Learning to manage global environmental risks (MIT, 2001), and Environment magazine. He co-chaired the recent study by the US National Research Council on Our Common Journey: A Transition Toward Sustainability. Clark is a member of the US National Academy of Sciences, and a recipient of the MacArthur Prize, the Humboldt Prize, and the Kennedy School's Carballo Award for excellence in teaching.

JOHN C. DERNBACH is a Professor of Law at Widener University and the former director of the Pennsylvania Department of Environmental Protection's (DEP) Office of Policy. DEP's Office of Policy identifies key issues, coordinates the development of regulations and policy initiatives, tracks and reviews all proposed rulemakings, and helps to provide long-range direction on a broad range of departmental goals and objectives. From 1981 to 1993, Dernbach held various posts with DEP's precursor, the Department of Environmental Resources, wrapping up his tenure there as director of the Advanced Science and Research Team. Dernbach also has been a professor at Widener University Law School since 1993, teaching classes in environmental law, international environmental law, property and administrative law and conducting seminars on global warming and sustainability. He also has extensive international environmental law experience, serving as a visiting lecturer at the University of Geneva's Graduate Institute of International Studies in Geneva, Switzerland, in 2003; at Macquarie University Law School in Sydney, Australia, in 1999; and at the University of Nairobi Law School in Nairobi, Kenya, in 1996. He is the editor of Stumbling Toward Sustainability (Environmental Law Institute, 2002), a comprehensive assessment of U.S. sustainable development efforts over the past decade. Dernbach attended the University of Wisconsin-Eau Claire, graduating summa cum laude and receiving his bachelor's degree in science in 1975. Dernbach received his juris doctor from the University of Michigan Law School in 1978, graduating cum laude.

SAM DRYDEN is a Managing Director of Wolfensohn & Company, a corporate advisory and investment firm located in New York, where he focuses on private equity investments in biofuels and other alternative energies. He is also CEO of

Emergent Genetics, LLC. Until June 2006, Sam served as the Chair and Corporate CEO of Emergent Genetics, Inc. The majority of the Company was acquired in April 2005 by the Monsanto Company and its remaining operations were acquired in June 2006 by Syngenta AG. Sam began his career as an Analyst with the US Department of Commerce's Bureau of Economic Analysis, with responsibilities for modeling and forecasting selected sectors of the US economy. He was then employed by the Union Carbide Corporation from 1974 to 1980. In 1980, Sam led the spin-out of Union Carbide's biotechnologies and related business operations and was subsequently co-founder, President and CEO of Agrigenetics Corporation. Sam was also chairman of an affiliated partnership which managed and invested $60 million in proprietary plant sciences research conducted in leading universities, as well as private and public research institutions worldwide. Sam founded and was President of Big Stone Inc.—a private venture-investment and development company focused on the life sciences. Sam also served as the non-executive chairman of Celgro Inc. He is a member of the Board of Directors of the Global Crop Diversity Trust. He has been an advisor to the Rockefeller, McKnight and MacArthur Foundations and a member of the Design Advisory Committee and Scientific Advisory Board of its African Agricultural Technology Foundation. Sam is a member of the Council on Foreign Relations and serves on its Advisory Committee on Intellectual Property and American Competitiveness. He has written and lectured widely on the policy issues of food security, the evolving nature of global public goods and new mechanisms for public and private sector relations. In this regard, his travels have taken him on missions to most countries in Latin America, including Cuba, as well as Europe, Asia, Africa, and the Middle East. Sam, a native of eastern Kentucky, received his B.A. degree in economics from Emory University in 1973.

KATHRYN FULLER is the Chair of the Board of Trustees, Ford Foundation and the former President and CEO of the World Wildlife Fund. Trained as both a lawyer and a biologist, Fuller took over the helm of the world's largest international conservation organization in 1989 after seven years serving first as director of WWF's wildlife trade monitoring program, then general counsel and executive vice president. Prior to joining WWF, she headed the Wildlife and Marine Resources Section of the Justice Department's Land and Natural Resources Division. At WWF, Fuller's emphasis has been on innovative conservation methods such as debt-for-nature swaps, conservation trusts, the inclusion of women in grass roots projects and creative partnerships to conduct conservation on large, eco-regional scales. Examples of large-scale projects undertaken during Fuller's tenure include creation of the world's first conservation trust fund for Bhutan and a partnership with the World Bank and the government of Brazil to triple the amount of rainforest under strict protection in the Amazon. In her 15 years as president and CEO, WWF has also doubled its membership, tripled its revenue and expanded its presence around the globe. She is a recipient of the

U.N. Environment Programme's Global 500 award and holds several honorary doctorates. Ms. Fuller also chairs the board of trustees of the Ford Foundation and sits on several other non-profit and corporate boards. She is a trustee of Brown University and a member of the Council on Foreign Relations. Fuller received her B.A. from Brown University and did graduate studies in marine, estuarine and environmental science at the University of Maryland.

GEORGE M. GRAY is the Assistant Administrator of the Office of Research and Development at US EPA. Prior to this position, Gray was Executive Director of the Harvard University Center for Risk Analysis (HCRA) and a lecturer at the University's Department of Health Policy and Management. Gray's primary research interests at Harvard included the characterization and communication of risk with a focus on food safety, agriculture, and chemicals in the environment. In addition to his post at HCRA, Gray serves on advisory committees for the Society of Toxicology, the Center for Food Safety and Applied Nutrition, the Food and Drug Administration and the National Institute for Environmental Health Sciences. He is also a member of the Environmental Literacy Council. Dr. Gray holds a B.S. in biology from the University of Michigan and a M.S. and Ph.D. in toxicology from the University of Rochester School of Medicine and Dentistry.

F. HENRY "HANK" HABICHT II currently serves as Managing Partner of SAIL Venture Partners, a leading venture capital fund investing in leading-edge clean energy, water, and related technologies. Prior to his SAIL affiliation, Mr. Habicht served as CEO of the Global Environment & Technology Foundation (GETF), where he now serves as Vice Chairman. He is a founding Principal of Capital E, LLC. Previously, Mr. Habicht was Senior Vice President of Safety-Kleen Corporation, a provider of industrial and recycling services to 400,000 customers with sales of over $1 billion. Prior to his position with Safety-Kleen, Mr. Habicht was Chief Operating Officer of U.S. EPA under Administrator William K. Reilly. Mr. Habicht initiated quality-oriented management improvements to improve planning and integrate U.S. EPA's diverse science, policy and enforcement functions. In addition, Mr. Habicht chaired or served on several interagency work groups concerning risk assessment, energy, transportation, trade, and technology promotion. From 1987 to 1989 Mr. Habicht was with William D. Ruckelshaus Associates as Vice President and Counsel. Prior to this position, Mr. Habicht was Assistant Attorney General of the United States where he directed the Land and Natural Resources Division with responsibility for all federal environmental enforcement, energy and natural resource litigation. Mr. Habicht is a member of numerous boards and advisory councils. He has served as a Member of the Secretary of Energy Advisory Board; and is currently on the Steering Committee of the Energy Future Coalition; Chairman of Board of Resolve, Inc.; Director of 3E Company; and as a Member of NREL National Advisory Board; and the President's Advisory Committee on Trade Policy and Negotiation; and the

Advisory Board for the National Leadership Summits for a Sustainable America. He also serves on the Dow Chemical Corporate Environmental Advisory Council, and the Princeton Environmental Institute and the National Pollution Prevention Roundtable Advisory Boards. Hank received a J.D. at the University of Virginia and A.B. at Princeton University.

JEREMY HARRIS recently completed his second and final term as Mayor of the City and County of Honolulu, Hawaii. As a trained environmentalist, Mayor Harris deeply appreciates the effects that government policy has on the world's ecosystems. Under his direction, the City and County of Honolulu has made great advances in curtailing urban sprawl, while preserving open space and prime agricultural lands. He has initiated numerous educational and volunteer programs to fight water pollution, preserve Honolulu's waters, and to recycle everything from paper and glass, to asphalt and construction materials. Mayor Harris is intensively involved in the multi-billion-dollar renovation of the Island's sewer system, as well as in a partnership with private industry to reclaim wastewater for agricultural and industrial uses. Before serving as Mayor, he worked for eight years as the City's Managing Director. In this capacity, Mayor Harris was a major force in the development of Honolulu's H-Power program, which uses trash as fuel to generate electricity. In its first six years of operation, the plant processed over four million tons of waste and generated electricity that would have otherwise would have required five million barrels of oil. Mayor Harris specialized in Marine Biology and Urban Ecosystems, obtaining his Master's Degree from the University of California at Irvine.

ROSALYN S. HOBSON, Associate Dean for Graduate Studies, School of Engineering at Virginia Commonwealth University, is an educator, researcher, and engineer. Her research interests include artificial neural networks and their application to control problems, intelligent systems, biological modeling and signal processing. As one of the founding faculty members of the School of Engineering at Virginia Commonwealth University, she has been instrumental in the establishment and success of the engineering program. She has established a partnership program between VCU and schools in South Africa and is designing a new engineering education program focused on engineering challenges in developing countries. She has served on and chaired numerous committees and developed a research group and laboratory in which projects on neural network applications are conducted. She has been awarded grants from industry and government, directed the research of numerous undergraduate and graduate students, published several articles and has been an invited lecturer in many venues. Additionally she was awarded a AAAS science and diplomacy fellowship to serve at the US Agency for International Development. Officially posted in the USAID Office of Education, Dr. Hobson served as the principal liaison between USAID and the National Academies for a study examining science and technology in US development assistance programs. She also served as a consultant for Invensys Control Systems. Dr. Hobson continues her work with numerous Richmond high schools and with outreach and summer programs that focus on promoting interest in and recruiting students into the engineering profession. Dr. Hobson's awards and honors include: the National Society of Black Engineers' Patricia A Lumpkin Educator of the Year Award, and the Frontiers in Education New Faculty Fellow. She served on the National Academy of Engineering Committee on Engineering Education and was selected for participation in the Stanford University National Science Foundation New Century Scholars Workshop. She has been nominated for the American Biographical Institute 1000 World Leaders of Scientific Influence and Who's Who of American Women 2000. Dr. Hobson received her B.S., M.S., and Ph.D. (1997) in Electrical Engineering from the University of Virginia.

JACK A. KAYE is the Director of the Research and Analysis Program of the Earth-Sun System Division in NASA's Science Mission Directorate. In this position, he has responsibility for the broad range of science research carried out in Earth and solar science at NASA centers, academia, other government agencies, and the private sector. Prior to being assigned to this position, he worked as a research scientist, program manager for atmospheric chemistry, and Research Division director for the former Office of Earth Science over his 21 year career at NASA. He represents NASA in numerous interagency activities related to climate, oceans, and Earth observations, and he serves as a member of the Steering Committee for the Global Climate Observing System. Among his many awards is his recognition in 2004 as a Meritorious Senior Executive. He is trained in chemistry, having received a B.A. from Adelphi University and a Ph.D. from the California Institute of Technology.

GERALD T. KEUSCH (IOM) is Director of the Global Health Initiative, Assistant Provost of the Medical Campus, and Associate Dean of the School of Public Health at Boston University. Prior to joining the university, Dr. Keusch was the Associate Director for International Research at the National Institutes of Health, and Director of the Fogarty International Center. He has been involved in clinical medicine, teaching and research for his entire career. Dr. Keusch's research has ranged from the molecular pathogenesis of tropical infectious diseases to field research in nutrition, immunology, host susceptibility, and the treatment of tropical infectious diseases and HIV/AIDS. He is involved in international health research and policy issues within the NIH, the Institute of Medicine, and the World Health Organization. Under his leadership, the programs of the Fogarty International Center have greatly expanded to address not only the pressing global issues in infectious diseases and the growing burden of non-communicable diseases, but also the critical cross-cutting issues such as the ethical conduct of research, intellectual property rights and global public goods, stigma, and the impact of improved health on economic development. Dr. Keusch is a graduate of Columbia College and Harvard Medical School, and he is Board Certified in Internal Medicine and Infectious Diseases.

KAI LEE is the Program Officer in the Conservation & Science Program at the David and Lucille Packard Foundation and former Rosenburg Professor of Environmental Studies at Williams College. He served as director of the Center for Environmental Studies at Williams College from 1991-1998 and is interim director in 2001-2002. Lee has continued to teach and conduct research on the relationship between technological change and democratic governance and is currently undertaking a study of urban sustainability. He was chair of the Committee on Long-Term Institutional Management of DOE Legacy Waste Sites at the National Research Council, 2001-2003. He serves now on the Water Science and Technology Board at the National Academies. He served on the National Research Council's Board on Sustainable Development and altogether Lee has served on eleven committees of the National Academies of the National Research Council: the Environmental Studies Board (1980-1982), the Board on Radioactive Waste Management (1983-88), the Committee to Assess Safety and Technical Issues at Department of Energy Reactors (1986-1987), the mitigation sub-panel of the Panel on Policy Implications of Greenhouse Warming (1989-1991), the Committee on Environmental Research (1991-1993), the Committee on Protection and Management of Pacific Northwest Anadromous Salmonids (1993-1995), the Board on Environmental Studies and Toxicology (1993-1995); the Board on Sustainable Development (1995-99); the Commission on Geosciences, Environment, and Resources (1996-99); the Committee on Long-Term Institutional Management of DOE Legacy Waste Sites: Phase 2 (2001-2003), and the Water Science and Technology Board (2004-2007). Additionally, in 1989 Lee was a member of the United Nations Environment Programme committee reviewing economic aspects of the Montreal Protocol on Substances that Deplete the Ozone Layer. Lee was educated in experimental physics at Columbia (A.B. magna cum laude, 1966) and Princeton (Ph.D., 1971).

J. TODD MITCHELL has served as chairman of HARC's board of directors since 2000 and was president of HARC from 2001 to 2006. Mr. Mitchell has a BA in Geology from The Colorado College (1981), and an MA in Geology from The University of Texas at Austin (1987). He has worked extensively in the energy industry, first as a co-founder of Strand Energy, an oil and gas exploration company, and later as co-founder of Rock Solid Images, a developer of seismic and petrophysical tools for reservoir characterization and imaging. Mr. Mitchell served on the board of directors of Mitchell Energy & Development Corp. for seven years, and is currently a director of Devon Energy, one of the country's largest producers of natural gas. From September 2006 to 2007, Mr. Mitchell was enrolled at the Centre for the Study of Environmental Change and Sustainability (Edinburgh, Scotland), where he focused on clean energy technology.

MARK MYERS is the Director of the U.S. Geological Survey. He is an internationally recognized geologist and former State Geologist and head of Alaska's Geological Survey. Mr. Myers, an expert on North Slope sedimentary and petroleum geology, served as survey chief for field programs in the MacKenzie Delta (ARCO, 1985), Cook Inlet (State of Alaska/U.S. Geological Survey, 1997), and North Slope (ARCO, 1999). He also served as sedimentologist for 13 other North Slope field programs. Mr. Myers is a past president and board member of the Alaska Geological Society; a certified professional geologist with the American Institute of Professional Geologists; a certified petroleum geologist with the American Association of Petroleum Geologists; and a licensed geologist with the State of Alaska. He served as an officer in the U.S. Air Force Reserve from 1977 to 2003, retiring as a Lt. Colonel. Mr. Myers received his doctorate in geology from the University of Alaska-Fairbanks in 1994, specializing in sedimentology, clastic depositional environments, surface and subsurface sequence analysis, and sandstone petrography. He earned his B.S. and M.S. degrees in geology from the University of Wisconsin-Madison.

RAY ORBACH is Under Secretary-Designate for Science at the Department of Energy (DOE). Prior to his nomination, Dr. Orbach served as Director of DOE's Office of Science. In this capacity, Dr. Orbach has managed an organization that is the third largest Federal sponsor of basic research in the United States, the primary supporter of the physical sciences in the U.S., and among the premier science organizations in the world. The Office of Science fiscal year 2006 budget of $3.6 billion funds programs in high energy and nuclear physics, basic energy sciences, magnetic fusion energy, biological and environmental research, and computational science. The Office of Science also provides management oversight of 10 DOE non-weapons laboratories, supports researchers at more than 275 colleges and universities nationwide, and builds and operates a suite of scientific facilities and instruments used annually by more than 19,000 researchers to extend the frontiers of all areas of science. From 1992 to 2002, Dr. Orbach served as Chancellor of the University of California (UC), Riverside. Under his leadership, UC Riverside doubled in size, achieved national and international recognition, and led the University of California in diversity and educational opportunity. Dr. Orbach's research in theoretical and experimental physics has resulted in the publication of more than 240 scientific articles. He has received numerous honors as a scholar including two Alfred P. Sloan Foundation Fellowships, the Joliot Curie Professorship at the Ecole Superieure de Physique et Chimie Industrielle de la Ville de Paris, the 1991-1992 Andrew Lawson Memorial Lecturer at UC Riverside, and the 2004 Arnold O. Beckman Lecturer in Science and Innovation at the

University of Illinois at Urbana-Champaign. Dr. Orbach received his Bachelor of Science degree in Physics from the California Institute of Technology in 1956. He received his Ph.D. degree in Physics from the University of California, Berkeley, in 1960 and was elected to Phi Beta Kappa.

LARRY PAPAY (NAE) is a private consultant, recently retired as Sector Vice President for the Integrated Solutions Sector of SAIC. He was responsible for business dealing with the integration of technology in the energy, environment and information areas for a variety of governmental and commercial clients worldwide. Prior to joining SAIC, Dr. Papay served as the Senior Vice President and General Manger of Bechtel Technology and Consulting as well as Senior Vice President of Southern California Edison Company. In February 2004, Dr. Papay was appointed to the Homeland Security Science and Technology Committee by Dr. Charles E. McQueary, Under Secretary for Science and Technology of the Department of Homeland Security. Prior to joining the Homeland Security Science and Technology Advisory Committee, Dr. Papay served on the President's Council of Advisors on Science and Technology and the Secretary of Energy Advisory Board. He is a member of the National Academy of Engineering and the American Nuclear Society and was recently elected to the NAE Council. Other present and past committee memberships include The National Academies Coordinating Committee on Global Change, the Charles Stark Draper Prize Committee, and the Committee on Science Engineering and Public Policy. Papay received a B.S. in Physics from Fordham University, and both a M.S. and Sc.D. in Nuclear Engineering from the Massachusetts Institute of Technology.

MERLE D. PIERSON serves as Deputy Under Secretary for Research, Education, and Economics (REE) at the United States Department of Agriculture. He previously served as Deputy Under Secretary for Food Safety and Inspection Service (FSIS) and Acting Under Secretary for Food Safety. As Deputy Under Secretary, Pierson provides leadership to the four agencies that comprise the Research, Education and Economics mission area: the Agricultural Research Service; the Cooperative State Research, Education, and Extension Service; the Economic Research Service; and the National Agricultural Statistics Service. While serving in the area of food safety, he had responsibility for overseeing FSIS policies and programs and the U.S. Codex Alimentarius office. Prior to his USDA appointment, Pierson served as Professor of Food Microbiology and Safety at Virginia Polytechnic Institute and State University (Virginia Tech) from 1970 to 2005. Pierson is a Fellow of the Institute of Food Technologists and the American Association for the Advancement of Science. Pierson is internationally recognized for his work on food safety management, in particular, Hazard Analysis and Critical Control Points (HACCP) and research on the reduction and control of foodborne pathogens. He has authored or co-authored more than 150 articles and 7 books on food safety and quality in addition to presenting numerous workshops on HACCP and food safety management. Pierson received his B.S. in biochemistry from Iowa State University and M.S. and Ph.D. in food science from the University of Illinois.

PRABHU PINGALI is an economist and Director of the Division of Agricultural and Development Economics at the UN Food and Agriculture Organization (FAO) is the President of the International Association of Agricultural Economists (IAAE). Pingali was Vice-President of the IAAE from 1997-2000 and chairman of the program committee for the 24th International Conference of Agricultural Economists. Pingali has more than 20 years experience in assessing the extent and impact of technical change in developing country agriculture in Asia, Africa and Latin America. Before joining FAO, Pingali was Director of the Economic Program at the International Maize and Wheat Improvement Center (CIMMYT) in, Mexico, the International Rice Research Institute at Los Baños, Philippines and the World Bank's Agriculture and Rural Development Department. An Indian national, Pingali earned his doctorate in Economics at North Carolina State University in 1982 in Raleigh, North Carolina, in the United States.

PETER H. RAVEN (NAS) is the Director of the Missouri Botanical Garden and Engelmann Professor of Botany at Washington University. His research interests include evolution of the plant family Onagraceae, conservation biology, biodiversity, and biogeography. Dr. Raven is former president of the American Association for the Advancement of Science and served as a member of President Clinton's Committee of Advisors on Science and Technology. He is a member of the National Academy of Sciences, where he has served on numerous boards and committees, has been the Home Secretary, and is currently the Chair of the Division on Earth and Life Studies (DELS). Dr Raven received his Ph.D. from the University of California, Los Angeles.

ROBERT STEPHENS founded and served as President of the Multi-State Working Group on Environmental Performance (MSWG), a national coalition of representatives from government, business, non-governmental organizations, and academic institutions in the US working on transformative policies relating to the environment and sustainable development. Via his continued involvement with the MSWG, Dr. Stephens serves as the Secretariat to the Best Practice Network for Sustainable Development (BPN) for the United Nations Environment Program, Division of Technology, Industry, and Economics. Dr. Stephens retired in July 2004 from the California EPA after 30 years of service, most recently as Assistant Secretary for Environmental Management and Sustainability. In this position, Dr Stephens was responsible for the development and implementation of programs leading to environmental policy innovation and sustainability in California. Over his career, Dr. Stephens also served as Deputy Director of the Department of Toxic Substances Control for Science, Pollution Prevention, and

Technology and Chief of the Hazardous Materials Laboratory for the state of California. Dr. Stephens is the primary and/or co-author of some 60 articles and book chapters ranging from basic environmental science and risk assessment to public policy related to the environment and sustainability. Dr. Stephens holds a Ph.D. in Chemistry from the University of California and has held prior positions in industry and academia.

CHARLENE WALL manages the North American Eco-efficiency Analysis program for the BASF Group. She furthers BASF's position as a global leader by facilitating the integration of sustainable development into the North American businesses. In 1992, she joined BASF Corporation, and has held positions in Product Development, Process Design Engineering and Safety and Ecology. In addition, she is the first chairperson for the American Institute of Chemical Engineer's Center for Sustainable Technology Practices.

STAFF

MARTY PERREAULT serves as the director of the Roundtable on Science and Technology for Sustainability. The National Academies have established the Science and Technology for Sustainability Program (STS) in the division of Policy and Global Affairs to encourage the use of science and technology to achieve long term sustainable development—increasing incomes, improving public health, and sustaining critical natural systems. She was previously the program director for the National Academies' Keck Futures Initiative, a 15-year effort to catalyze interdisciplinary inquiry and to enhance communication among researchers, public and private funding organizations, universities, and the general public. Prior to joining the National Academies, Marty was the Vice President for Community Initiatives with the Orange County (CA) United Way. Before that she was the Manager of Community Services for the American Red Cross National Headquarters. She received her master's degree in Industrial Engineering with a focus on Health Systems from Rensselaer Polytechnic Institute.

PATRICIA KOSHEL is a senior program officer with the National Academies' Policy and Global Affairs Division. She has been the staff lead for a consensus study on science and technology in US Foreign Assistance Programs and has also worked on the Science and Technology for Sustainability Program. Before joining the National Academies, Pat was the Director of Bilateral Programs in the Office of International Affairs at the US Environmental Protection Agency. Before that she was the Energy and Environmental Policy Advisor for the US Agency for International Development. She has a master's degree in economics.

DEREK VOLLMER is the Program Associate for the Science and Technology for Sustainability Program (STS) at the National Academies. In this position, he supports a variety of program activities including the Roundtable on Science and Technology for Sustainability and a joint energy and air pollution between the NAE and the Chinese Academies. He has also helped develop a program on urban environmental sustainability and supported the "Strengthening Science-Based Decision Making" workshop series. He is concurrently pursuing his M.S. in Environmental Science and Policy at Johns Hopkins University. Before moving to Washington, Derek lived in Yanji, China, teaching English and music at a technical school as part of the Salesian Lay Missioners program. He graduated Summa Cum Laude from the University of Notre Dame in 2002 with a B.A. in Government and International Studies, where he wrote an honors thesis detailing China's role in global climate change. He speaks Mandarin Chinese and French.

KATHLEEN MCALLISTER is the Senior Program Assistant for the Science and Technology for Sustainability Program (STS) at the National Academies. Before joining The National Academies, she attended Lehigh University and graduated with highest honors as well as departmental honors in 2006 with a B.A. in Sociology. Kathleen wrote an honors thesis on social implications of human trafficking into the United States and worked throughout her college career as a Research Assistant for Professors of Sociology at Lehigh University. She also speaks conversational Spanish, and has had internships in the offices of U.S. Representative Paul E. Kanjorski and U.S. Senator Arlen Specter.

F

Federal Sustainability R&D Forum Workshop Participants

Richard B. Alexander (USGS Panelist)
Matt Arnold (Sustainable Finance Ltd.)
Ghassem Asrar (USDA Panelist)
Adela Backiel (USDA)
Sarah Banas (AAAS)
Donna Myers (USGS)
Ann Bartuska (USFS)
Diana Bauer (DOT)
Jacques Beaudry-Losique (DOE Panelist)
Bill Behn (House Committee on Science and Technology)
Mike Bertolucci (Interface Research Corporation)
Rich Bissell (NAS)
Evan Braneksy (WRI)
Robert Brock (NOAA)
Paul Brubaker (DOT)
Randy Bruins (EPA)
John Carberry (DuPont)
Leslie Carothers (Environmental Law Institute)
Steve Carpenter (University of Wisconsin, Madison)
Colleen Charles (USGS)
William Chernicoff (DOT Panelist)
Ralph Cicerone (NAS)
Bill Clark (Harvard University)
Greg Crosby (USDA/CSREES)
Patrick Davis (DOE)

Sam Dryden (Wolfenson & Co)
Cliff Duke (Ecological Society of America)
Micheal Eaton (Resources Legacy Fund)
Ned Euliss (USGS Panelist)
Jana Gastellum (UN Foundation)
Mary Glackin (NOAA)
James Fischer
Iris Goodman (CENR Ecosystems Services Working Group Speaker)
Tameka Gore (NOAA)
Linda Gundersen (USGS)
Hank Habicht (GETF)
Ed Hackett (NSF)
John Hall (DOD)
Alan Hecht (EPA)
Paul V. Dresler (USGS)
Rosalyn Hobson (Virginia Commonwealth University)
William Hohenstein (USDA)
Dan Kammen (UC Berkeley)
Brian Keinman (OMB)
Jerry Keusch (Boston University)
Pat Koshel (NAS)
Julia Kregenow (NAS)
Kristi Kubista-Hovis (OMB)
Dan Kugler (USDA Panelist)
Sarah LaPlante (USFS)
Kirsten Larsen (NOAA)
Kai Lee (Packard Foundation)
Audrey Levine (EPA)
Will Logan (NAS)
James Mahoney (NOAA, ret.)
Dale Manty (EPA)
Pam Matson (Stanford University)
Kathleen McAllister (NAS)
Todd Mitchell (Houston Advanced Research Center)
John Mizroch (DOE)
Steve Murawski (NOAA)
Mark Nechodom (USFS Panelist)
Dan Nees (WRI)
Dr. World Nieh (USFS)
Margaret Palmer (NSF Panelist)
Steve Parker (NAS)
Kara Parks (NOAA)
Marcia Patton-Mallory (USFS Panelist)

Donna Perla (EPA)
Marty Perreault (NAS)
Timothy Petty (Dept. of Interior)
Melanie Roberts (NSF AAAS Fellow)
Bruce Rodan (OSTP-Federal Policies Panelist)
Sara Scherr (EcoAgriculture Parters)
Dr. Jim Sedell (USFS)
Seth Shames (EcoAgriculture Partners)
Carl D. Shapiro (USGS)
Robbin Shoemaker (USDA)
Emmy Simmons (USAID) (retired)
Mike Sissenwine
Jeffrey J. Steiner (USDA)
Greg Symmes (NAS)
Larry Tieszen (USGS)
Woody Turner (NASA Panelist)
Michael Uhart (NOAA)
Nathalie Valette-Silver
Uday Varadarajan (DOE)
Derek Vollmer (NAS)
Dan Walker (OSTP)
Marca Weinberg (USDA)
Pamela Williams (EPA)
Rob Wolcott (EPA)

G

Biographical Information Federal Sustainability R&D Forum Speakers and Panelists

RICHARD B. ALEXANDER is a Research Hydrologist with the USGS NAWQA (National Water Quality Assessment) program and has been with the USGS for 27 years. His research focuses on the development and use of water-quality modeling techniques to investigate pollutant sources and contaminant transport processes in surface waters. He is a co-developer of the USGS SPARROW water-quality model. His studies include assessments of nutrient sources and processes in streams of the Mississippi River Basin and their influence on nutrient delivery to the Gulf of Mexico. He has also developed models of nutrients and pathogens in the surface waters of New Zealand as a visiting scientist with the National Institute of Water and Atmospheric Research. He is a coordinator and instructor for USGS technical courses on statistical methods and water-quality modeling and the Associate Editor for water-quality modeling for the Journal of the American Water Resources Association. His educational background includes a M.S. in Water Resources Administration from the University of Arizona and a B.A. from the University of North Carolina-Charlotte.

JACQUES BEAUDRY-LOSIQUE serves as the Program Manager of the U.S. Department of Energy (DOE)'s Office of Biomass Program. The Office leads federal efforts to develop technologies that will enable clean biofuels from abundant domestic resources to significantly reduce U.S. dependence on oil. Mr. Beaudry-Losique initially joined the Department as Manager of the Industrial Technologies Program (ITP) in June 2005, serving in that capacity until reappointed to the Office of Biomass Program (OBP) in December 2006. He brings to the Office extensive experience in executive management, business development and commercial negotiations. He was instrumental in modernizing ITP's portfolio and is positioning the accelerated growth of the Biomass program toward market results. Prior to joining DOE, he served two years as a mergers and acquisitions consultant, acting CFO and board member to many small and midsize technology companies. Before that he was the business development leader of General Electric Power Systems investment activities. There, he was responsible for the placement of more than $20 million in equity investments into strategic technology companies, and oversight of more than $75 million of GE investments. Prior to that, he devised growth strategies for Aspen Technologies, a leading engineering and supply chain software company. Mr. Beaudry-Losique also has many years of experience as a management consultant with McKinsey and Company. Mr. Beaudry-Losique holds a Bachelor of Science degree in chemical engineering from the University of Montreal and a Master of Science degree in Industrial Engineering and Engineering Management from Stanford University. As a recipient of a Canadian Science Foundation Fellowship, he attended the MIT Sloan School of Management, where he received a master's degree in management in 1992.

RANDY BRUINS is an environmental scientist in the U.S. EPA's Office of Research and Development. He received his bachelor's (1978) and master's (1980) degrees, both in Zoology, from Miami University and his Ph.D. (1997) in environmental science from Ohio State University. His dissertation research examined methods for reducing flooding in central China, through ecological strategies such as replacement of low-lying rice with native wetland crops. Since 1997 Randy's EPA research has focused on methods for integrating ecological risk assessment and economic analysis. He addressed these topics in a 2005 book, Economics and Ecological Risk Assessment: Applications to Watershed Management (co-edited with Matthew Heberling) and a forthcoming (2007) volume, Valuation of Ecological Resources: Integration of Ecology and Socioeconomics in Environmental Decision Making (co-edited with Ralph Stahl, Larry Kapustka and Wayne Munns). Randy serves on the board of the U.S. Society for Ecological Economics, and he is a coauthor of EPA's 2006 strategy for measuring the benefits of ecosystem protection (the Ecological Benefits Assessment Strategic Plan). His current position is in EPA's National Exposure Research Laboratory, where he is leading the Future Midwestern Landscapes Study, an examination of ecosystem services in the Midwestern US with special emphasis on the implications of biofuels development.

STEPHEN (STEVE) CARPENTER is the Stephen Alfred Forbes Professor of Zoology at the University of Wisconsin-Madison Center for Limnology. He directs the North Temperate Lakes Long-Term Ecological Research site as well as a diverse program of whole-ecosystem experiments. He is co-Editor in Chief of Ecosystems, and a member of governing boards for the Beijer Institute of Ecological Economics, Institute of Ecosystem Studies, and Resilience Alliance. Carpenter is a member of the U.S. National Academy of Sciences, a Fellow of the American Academy of Arts and Sciences, and a foreign member of the Royal Swedish Academy of Sciences. He has received many awards for distinguished research. Among these are a Pew Fellowship in Conservation and Environment, the G. Evelyn Hutchinson Medal of the American Society of Limnology and Oceanography, the Robert H. MacArthur Award from the Ecological Society of America, the Excellence in Ecology Prize for Limnetic Ecology, the Naumann-Thienemann medal of the International Society for Limnology, many honors from the U.W.-Madison campus, and election to the Ralf Yorque Society. The Institute for Scientific Information has recognized him as one of the world's most highly cited researchers in Environmental Science. From 2000-2005 he served as co-chair of the Scenarios Working Group of the Millennium Ecosystem Assessment. He served as President of the Ecological Society of America in 2000-2001. Carpenter is an ecosystem ecologist known for his leadership of large-scale experiments and adaptive ecosystem management. His work has addressed trophic cascades and their effects on production and nutrient cycling, contaminant cycles, freshwater fisheries, eutrophication, nonpoint pollution, ecological economics of freshwater, and resilience of social-ecological systems. Carpenter has published 4 books and about 300 scientific papers. He received a B.A. from Amherst College (1974), M.S. from University of Wisconsin-Madison (1976), and Ph.D. from U.W. Madison (1979). From 1979-1989 he served as Assistant and then Associate Professor at the University of Notre Dame. He joined the U.W.-Madison faculty in 1989.

WILLIAM CHERNICOFF is an engineer within the USDOT Research and Special Programs Administration He specializes in alternative fuel and clean / advanced vehicle propulsion technology and sustainable transportation, and is a lead specialist for DOT's hydrogen engineering activities. His primary focus is on vehicle and infrastructure safety and operations through the development and implementation of codes, standards, and best practices for vehicles and infrastructure. Additionally, he leads efforts for development and deployment of several advanced propulsion vehicles and infrastructure. As part of the USDOT work efforts focus on ensuring the operational safety, security, and reliability of the transportation system, and maintaining the public confidence in deployed technologies. Mr. Chernicoff is an active participant on several codes and standards committees for hydrogen, natural gas, and alternative fuels. He is a member of the USDOT Hydrogen Working Group and chairs the safety, codes and standards action team. He holds a BS in Materials Engineering from MIT, an MS in manufacturing engineering from Boston University, and is pursuing a PhD in Mechanical Engineering at Tufts University.

NED "CHIP" EULISS JR. obtained his M.S. from Humboldt State University in California and his Ph.D. from Oregon State University. Dr. Euliss has served as a Research Wildlife Biologist with the US Geological Survey's Northern Prairie

Wildlife Research Center in Jamestown, North Dakota since 1986. Dr. Euliss has led long-term research and monitoring of the dynamics in hydrology, chemistry, and biology of a prairie wetland complex at the Cottonwood Lake Study Area near Jamestown. Dr. Euliss has been instrumental in leading research on carbon sequestration, and ecosystem goods and services provided by glacial prairie wetlands. Currently, Dr. Euliss is the interdisciplinary team lead for the Integrated Landscape Monitoring—Prairie Pilot. This project aims to quantify, monitor, and model ecosystem goods and services across the Prairie Pothole Region of the U.S. across space and time. Dr. Euliss also serves on the Department of the Interior's task force on Climate Change. Dr. Euliss is professionally affiliated with the Society of Wetland Scientists, the American Association for the Advancement of Science, The Wildlife Society, and the North American Benthological Society.

JIM FISCHER is the Senior Energy Advisor in the Research, Education, and Economics Directorate at the U.S. Department of Agriculture. He holds a Ph.D. in agricultural engineering from the University of Missouri-Columbia. As a USDA research engineer in the 1970s, he published the design specifications for the original integrated on-farm energy system. Dr. Fischer has served at three universities (holding Professorial and Dean positions) — Missouri, Michigan State, and Clemson. He has provided leadership for numerous national organizations as well as led national programs envisioning the future of state and land-grant universities. His leadership in these organizations and programs has resulted in responsive research and outreach programs and relevant curricula at universities that address the critical issues impacting society today, such as agriculture, food, environment and energy. He has published more than 100 papers, contributed book chapters, testified before Congress, and served on peer review panels and advisory boards. In June 2003, he was appointed to the Board of Directors for the Energy Efficiency and Renewable Energy programs of the U.S. Department of Energy. As the Senior Technical Advisor (Academe) he developed innovative partnerships and models of collaboration with universities, especially land grant universities, US Department of Agriculture, foundations and the agricultural, industrial and business communities. In January 2007, he and his wife, Sharon, formed James R. Fischer and Associates; a company focused on technology and management issues at the intersection of agriculture and energy. Presently, this company is developing energy science and education programs for the U S Department of Agriculture and is also working with US Department of Energy in developing partnership with Land Grant Universities' for energy outreach and education programs.

IRIS GOODMAN is an environmental scientist in the US EPA's Office of Research and Development. She received her bachelor's (1980) from the University of Maryland, in Conservation and Resource Management, with a minor in Economics. She received her master's degree from the University of Wisconsin-

Madison (1987), in Water Resources Management. She served for three years as a science analyst for the Congressional Office of Technology Assessment, and two years staffing environmental committees of the Wisconsin state legislature. During her career at EPA, she has written federal rules to protect groundwater; worked as a Principal Investigator at the National Exposure Research Laboratory on issues related to hydrology and landscape ecology at regional scales; worked as a visiting scientist for the Interior Columbia Basin Ecosystem Management Project, a $40 million study on adaptive management led by USDA's Forest Service and DOI's Bureau of Land Management; served as a Regional Scientist in EPA's Region 10, Seattle; and managed the ecology extramural grants program for ORD's National Center for Environmental Research. Currently, Iris is acting Deputy to Rick Linthurst, EPA's National Program Director for Ecology. She also co-chairs, with Robert Doudrick, the Ecosystem Services Workgroup, convened under OSTP's Subcommittee on Ecological Systems.

DANIEL M. KAMMEN is the Class of 1935 Distinguished Professor of Energy at the University of California, Berkeley, where he holds appointments in the Energy and Resources Group, the Goldman School of Public Policy, and the department of Nuclear Engineering. Kammen is the founding director of the Renewable and Appropriate Energy Laboratory (RAEL). Kammen is also the Co-Director of the Berkeley Institute of the Environment. Kammen received his undergraduate (Cornell A., B. 1984) and graduate (Harvard M. A. 1986, Ph.D. 1988) training is in physics After postdoctoral work at Caltech and Harvard, Kammen was professor and Chair of the Science, Technology and Environmental Policy at Princeton University in the Woodrow Wilson School of Public and International Affairs from 1993–1998. Through RAEL, Kammen works with faculty colleagues, postdoctoral fellows, and roughly 20 doctoral students on a wide range of science, engineering, economics and policy projects related to energy science, engineering and the environment. The focus of Kammen's work is on the science and policy of clean, renewable energy systems, energy efficiency, the role of energy in national energy policy, international climate debates, and the use and impacts of energy sources and technologies on development, particularly in Africa and Latin America. His work is interdisciplinary, and extends from theoretical studies to highly practical field projects and the design and development of specific policy initiatives and pieces of legislation. Kammen has published five books, over 200 journal articles and 30 research reports. Daniel Kammen serves on the National Advisory Board of the Union of Concerned Scientists, on the Technical Review Board of the Global Environment Facility is on the advisory board of the Union of Concerned Scientists, and in 1998 was elected a Permanent Fellow of the African Academy of Sciences. In February 2007, Kammen received the Distinguished Citizen Award from the Commonwealth Club of California. In February, 2007, Kammen received the Distinguished Citizen Award from the Commonwealth Club of California.

DANIEL E. KUGLER is the Deputy Administrator for Natural Resources and Environment at the Cooperative State Research, Education, and Extension Service, U.S. Department of Agriculture, in Washington, DC. Dan is a member of the Senior Executive Service and provides leadership and administration to programs and issues in water quality and availability, soil/land and air resources; forest resources and products; global change; ecology; sustainable development; conservation; sustainable natural resource management; environmental quality; rangeland resources; and wildlife and fisheries. Dan served as a math-science teacher training Peace Corps volunteer in Afghanistan from 1971-1973. From 1976 to 1986, he was an agricultural economist with USDA Economic Research Service working on soil depletion economics and policy, and an agricultural sector assessment for the Syrian Arab Republic. Dan joined the former Cooperative State Research Service in 1986 where he became Deputy Administrator for Special Programs, leading and administering programs in agricultural industrial materials, aquaculture, small scale agriculture, and sustainable agriculture. From 1995 to 1999, Dan was Section Leader for Processing, Engineering, and Technology in the Plant and Animal Systems unit of CSREES, where he provided leadership for agricultural engineering, small farms, biobased products, food safety and science, and farm safety. From 1999 to 2002, he was Deputy Administrator for Economic and Community Systems, providing leadership for place-based rural and community prosperity and development, sustainable agriculture, urban agriculture, diversity, financial security, risk management education, small farms, digital access and literacy, and entrepreneurialism. Dan has a B.S. in Physics, M.S. in Resource Development, and Ph.D. in Agricultural Economics from Michigan State University.

MARCIA PATTON-MALLORY currently works in the office of the Chief, USDA Forest Service, as the Biomass and Bioenergy Coordinator. Previous to this assignment, she was the Director of the Rocky Mountain Research Station of the USDA Forest Service, in Fort Collins, CO for five years, and Assistant Station Director for 10 years. Additional assignments with the USDA Forest Service include a Research Staff Specialists in the Washington Office and a Research Engineer at the Forest Products Laboratory in Madison, WI. Marcia's technical training includes a B.S. in Wood Science and Technology from the College of Forestry and Natural Resources at Colorado State University, and M.S. and Ph.D. in Civil Engineering from the College of Engineering at Colorado State University. Marcia's special assignments include a Science and Technology Fellow, assigned to the U.S. Senate.

JOHN MIZROCH is the Principal Deputy Assistant Secretary in Office of Energy Efficiency and Renewable Energy at the US Department of Energy. John joined the Department of Energy from his previous position as President and

CEO of the World Environment Center (WEC). At the WEC, he worked to advance sustainable development by encouraging environmental leadership, helping improve health and safety practices worldwide, and fostering the efficient use of natural resources to protect the global environment. Prior to leading the WEC, Mizroch promoted environmental technology transfer and investment in the developing world including Latin America, Asia, and Eastern Europe. Mizroch has also been a member of the Trade and Environmental Policy Advisory Committee at the U.S. Trade Representative's Office and also served on the Cleaner Fossil Fuel Systems Advisory Committee of the World Energy Council. Mizroch, an attorney, has served as a Foreign Service officer in South Africa, a senior official at the U.S. Department of Commerce in the Reagan and Bush administrations, and as a senior advisor to the Joint Economic Committee of the U.S. Congress. He received undergraduate and graduate degrees from the University of Virginia and a law degree from The College of William and Mary in Williamsburg, Virginia. The U.S. Department of Energy's Office of Energy Efficiency and Renewable Energy invests in a diverse portfolio of energy technologies to provide efficient, clean and renewable energy leading toward a stronger economy, a cleaner environment, and greater energy independence for America.

MARK NECHODOM is a Research Social Scientist with the Pacific Southwest Research Station (U.S. Forest Service), and lead scientist for the Social and Policy Sciences on the Sierra Nevada Framework Science Team. He led the socioeconomic and institutional analysis team for the Lake Tahoe Watershed Assessment (published March 2000) and is an author of the Adaptive Management Strategy for the Lake Tahoe Environmental Improvement Program. Mark has been with the Forest Service since September 1998. Mark holds a Ph.D. in political science from the University of California, Santa Cruz, with an emphasis in geography and democratic political theory.

MARGARET PALMER is Professor and Director of the Chesapeake Biological Laboratory (CBL), University of Maryland Center for Environmental Science. Dr. Palmer graduated Phi Beta Kappa from Emory University in 1977, then went on to earn her Ph.D. from the University of South Carolina in coastal oceanography in 1983. Palmer's research expertise is riverine science and coastal linkages, particularly human interactions with land and water. She has more than 100 scientific publications, serves as an editor for the journal Restoration Ecology and published the book The Foundations of Restoration Ecology in 2006. Dr. Palmer has been honored as a AAAS Fellow, an Aldo Leopold Leadership Fellow, a Lilly Fellow, a Distinguished Scholar Teacher, and with an Ecological Society of America Distinguished Service Award. Dr. Palmer serves on boards for the Chesapeake Bay Trust, American Rivers, the NSF National Center for Earth Surface Dynamics and the NSF Long Term Ecological Research program.

STEVE PARKER is a senior staff member at the National Research Council. He is Director of the Water Science and Technology Board (since 1982). From 1990 until early in 1997, he also served concurrently as Associate Executive Director of the Commission on Geosciences, Environment, and Resources. From 1997-2000, he served concurrently as Director of the Board on Natural Disasters. With the WSTB, Parker is responsible for study programs in a broad range of water related and natural resources topics. Subject areas include aquatic ecology and restoration; ground water science, technology, and management; hydrologic science; water quality and water resources management; pollution control; and other related topics. Some recent, selected publications on which he did principal NRC staff work include Hydrologic Science Priorities for the US Global Change Research Program (2000), Watershed Research in the U.S. Geological Survey (1997), Hazardous Materials in the Hydrologic Environment (1996), Mexico City's Water Supply (1995), National Water Quality Assessment: The Challenge of National Synthesis (1994), and Opportunities in the Hydrologic Sciences (1991). Parker technical expertise lies principally in hydrologic engineering and water resources systems analysis. Prior to joining the NRC in 1982, he was in charge of river basin planning studies at the Federal Energy Regulatory Commission (1979-1982). From 1972-1979, he was with the New England Division of the Army Corps of Engineers, where he was acting chief of hydrologic engineering; the focus of his technical work included water quality, flood and drought, and hydropower system studies. From 1970-1972, Parker was employed by Anderson-Nichols consulting engineers in Boston where he worked on water supply oriented projects. In 1969-1970, Parker served in the U.S. Navy in Vietnam, where he commanded a river patrol (Swift) boat. Parker was educated in hydrology and civil engineering at the University of New Hampshire (B.S.) and did graduate work in hydrology and business administration. He is a certified Professional Hydrologist, a member of the research advisory board of the National Water Research Institute, and a member of the American Institute of Hydrology and American Water Resources Association.

BRUCE D. RODAN a Senior Policy Advisor-Environment in the White House Office of Science and Technology Policy (OSTP). Dr. Rodan serves as OSTP liaison to the Ecosystems and the Toxics and Risk Subcommittees of the NSTC Committee on Environment and Natural Resources (CENR). Dr. Rodan is a medical doctor (U. Melb) with Masters Degrees in Environmental Studies (U. Melb) and Public Health (Harvard). His work has included environmental risk analyses for toxic chemicals under the U.S. EPA Integrated Risk Information System (IRIS), negotiating the Stockholm Convention on Persistent Organic Pollutants (POPs), and research on neotropical timber species under the CITES Treaty.

SARA J. SCHERR is an agricultural and natural resource economist specializing in land and forest management policy in tropical developing countries. She is Director of Ecoagriculture Partners, an international partnership to promote increased productivity jointly with enhanced natural biodiversity and ecosystem services in agricultural landscapes. She also serves as Director of Ecosystem Services for Forest Trends, an NGO that promotes forest conservation through improved markets for forest products and ecosystem services. She is a member of the United Nations Millennium Project Task Force on Hunger, and a member of the Board of Directors of the World Agroforestry Centre. Dr. Scherr's previous positions include: Adjunct Professor at the University of Maryland, College Park, USA; Co-Leader of the CGIAR Gender Program; Senior Research Fellow at the International Food Policy Research Institute in Washington, D.C.; and Principal Researcher at the World Agroforestry Centre, in Nairobi, Kenya. She was previously a Fulbright Scholar (1976), and a Rockefeller Social Science Fellow (1985-1987). Dr. Scherr received her B.A. in Economics at Wellesley College in Massachusetts, and her M.Sc. and Ph.D. in International Economics and Development at Cornell University in New York.

JEFFERY STEINER is the USDA Agricultural Research Service National Program Leader for Agricultural System Competitiveness and Sustainability. The research mission of this program is to develop integrated technology and information solutions that solve problems related to agricultural productivity, profitability, energy efficiency, and natural resource stewardship for different kinds and sizes of farms. Dr. Steiner joined the ARS National Program Staff in January 2006 after 17 years as a Research Agronomist at the ARS National Forage Seed Production Research Center in Corvallis, Oregon. He received his B.S. and M.S. in Plant Science from California State University-Fresno, and the Ph.D. from Oregon State University. Prior to joining ARS in 1988, he was an Associate Professor in the Plant Science Department at CSU-Fresno. Dr. Steiner is also a Fellow of the American Society of Agronomy and Crop Science Society of America, and is the 2007 recipient of the CSSA Seed Science Award.

WOODY TURNER is the Program Scientist for Biological Diversity and Program Manager for Ecological Forecasting in the NASA Headquarters Science Mission Directorate. As program scientist, he oversees the agency's basic research efforts to use satellite-derived information to understand the relationship of biodiversity to climate, landscape change, and ecosystem function. The NASA Ecological Forecasting Program is an applications activity seeking to bring together satellite observations and ecological models to support decision making for conservation biology and sustainable regional development. Born in Nashville, TN, Woody graduated from the University of North Carolina, Chapel Hill in 1982 and earned master's degrees in public affairs from Princeton University in 1987 and in conservation biology from the University of Maryland in 2001. He lives in Chevy Chase, Maryland, with his wife Jennifer and their two children.

DAN WALKER is Senior Policy Analyst in the Office of Science in the Office of Science and Technology Policy (OSTP). He previously held the title of senior program officer at the Ocean Studies Board at The National Academy of Sciences. Since 1999, Dr. Walker has held a joint appointment as a Guest Investigator at the Marine Policy Center of the Woods Hole Oceanographic Institution. He received his Ph.D. in Geology from the University of Tennessee in 1990. Dr. Walker has directed a number of NRC studies including Clean Coastal Waters: Understanding and Reducing the Effects of Nutrient Pollution (2000), Science for Decisionmaking: Coastal and Marine Geology at the U.S. Geological Survey (1999), Global Ocean Sciences: Toward an Integrated Approach (1998), and The Global Ocean Observing System: Users, Benefits, and Priorities (1997). A former member of both the Kentucky and North Carolina State geologic surveys, Dr. Walker's interests focus on the value of environmental information for policymaking at local, state, and national levels.

STEERING COMMITTEE

ANN BARTUSKA became the Deputy Chief for Research & Development in January 2004. Dr. Bartuska is a nationally and internationally recognized leader in natural resource science and management. Her professional activities have been as Program Manager for the National Acid Precipitation Program at North Carolina State University (1982-1987) then for the US Forest Service (1987-1989), Assistant Director for the US FS Southeastern Forest Experiment Station, Director of Forest Health Protection (1994-1999), Director of Forest and Range Management (1999-2001), and Executive Director, Invasive Species Initiative, The Nature Conservancy. She was elected President of the Ecological Society of America (2003) and served on the Board of the Council of Science Society Presidents.

GREG CROSBY is the Agency Representative to the USDA Council for Sustainable Development; Governing Committee and Directors Council for the Cooperative Extension eXtension (e-Extension on-line learning) Initiative; US Government and USDA Lead for WSSD Partnership on Geographic Learning for Sustainable Development; Director of Agency One Solution (on-line reporting) Initiative; Director of Geospatial Extension Specialists NRI Program; Agency Working Group on Science for Sustainability; Co-chair of the Cooperative Extension Service Workforce Development Initiative; and Senior eGovernment Fellow at the Council for Excellence in Government. His career experiences include building local alliances for science and technology education, Triangle Coalition for Science and Technology Education (NSTA) and Carnegie Corporation; Curriculum writer for ChemCom: Chemistry in the Community, American Chemical Society and University of Maryland; Aerospace Education Specialist,

NASA Goddard Space Flight Center; Research Associate, National Academy of Sciences, NRC.

LINDA GUNDERSEN is the Chief Scientist for Geology at the U.S. Geological Survey. She has worked as a geologist and senior manager with the USGS for over 28 years. The first half of her career focused on conducting research and directing diverse projects in the fields of geochemistry and ore deposits, ranging from understanding the origin of hard rock uranium deposits to environmental investigations of radon, uranium, and radium in rocks, soils, and water; eventually assessing the geologic radon potential of the United States. This research was conducted in partnership with states, federal agencies, universities, and communites, and received significant grants from the Department of Energy and the Environmental Protection Agency. From 1994 to 1998 she served as Program Coordinator of the Energy Resources Program and the Mineral Resources Program. As a member of the National Academy of Sciences Committee on Risk Assessment of Exposure to Radon in Drinking Water from 1997-1999, she participated in the first comprehensive overview and revamping of our understanding of this risk. In 1998, she became the Associate Chief Geologist for Operations implementing successful changes in the budgeting and science planning processes of the Geologic Division. During 2000 she served in the Director's Office, working with the Director and Deputy Director in leading and implementing a major reorganization of the USGS management and science planning structure. She was appointed Chief Scientist for Geology in 2001, supervising the geologic hazard, resource, and landscape programs of the USGS. From 2005-2006 she served on detail as the Associate Director for Geology. Currently she is a champion for data interoperability across the geosciences, organizing and developing a Geosciences Information Network with the State Geological Surveys. Her academic background includes undergraduate and graduate work in structural geology and geochemistry at the State University of New York at Stony Brook and at the University of Colorado, Boulder. She has received the Department of Interior Superior Service and Meritorious Service Awards, the Unit of Excellence Award, and the Secretary of the Interior's Bronze Executive Leadership award. She has published over 65 papers in the field of geology.

ALAN HECHT is the Director for Sustainable Development in the U.S. Environmental Protection Agency's Office of Research and Development. He served as Director for International Affairs at the National Security Council and Associate Director for Sustainable Development at the White House Council on Environmental Quality (2002-2003). Hecht participated in both the 1992 Earth Summit in Rio and the 2003 World Summit on Sustainable Development in Johannesburg. His most recent publication on sustainable development is in the Environmental Law Institute's magazine The Environmental Forum (September/October 2003). Spanning a federal career of 29 years, Hecht previously served as the Principal

Deputy and Deputy Assistant Administrator for International Activities at the U.S. EPA (1989-2001). He was the Acting Assistant Administrator for International Activities from 1992-1994. Hecht served in science and policy positions with the National Oceanographic Administration (1982-1989) and the National Science Foundation (1976-1982). He was Director of the National Climate Program from 1981 to 1989, and Director of the Climate Dynamics Program at NSF from 1976 to 1981. He has published numerous technical reports, edited two books on paleoclimatology and served as Chief Editor for journals of the American Meteorological Society.

STEVE MURAWSKI serves as Director of Scientific Programs and Chief Science Advisor for NOAA Fisheries. He is responsible for about 30 laboratories, eight offshore research vessels, and 1,400 staff throughout the United States. His organization's mission is to provide the scientific basis for conservation and management of living marine resources and their ecosystems. Dr. Murawski was previously the Director of the Office of Science and Technology, a position he held since October 2004. Prior to coming to NOAA Fisheries headquarters, he served as Chief Stock Assessment Scientist for the Northeast Fisheries Science Center in Woods Hole, Massachusetts (1990-2004). His research background is in fisheries biology and stock assessment. His current roles include official U.S. delegate to the International Council for the Exploration of the Sea, member on the Global Ocean Ecosystems Dynamics (GLOBEC) Program Steering Committee, and Project Manager for NOAA Fisheries' Ecosystem Management Pilot Projects.